COMPLEXITY

Other interview books from Automatic Press ♦ $\frac{\vee}{\mathsf{I}}$P

Formal Philosophy
edited by Vincent F. Hendricks & John Symons
November 2005

Masses of Formal Philosophy
edited by Vincent F. Hendricks & John Symons
October 2006

Political Questions: 5 Questions for Political Philosophers
edited by Morten Ebbe Juul Nielsen
December 2006

Philosophy of Technology: 5 Questions
edited by Jan-Kyrre Berg Olsen & Evan Selinger
February 2007

Game Theory: 5 Questions
edited by Vincent F. Hendricks & Pelle Guldborg Hansen
April 2007

Legal Philosophy: 5 Questions
edited by Morten Ebbe Juul Nielsen
October 2007

Philosophy of Mathematics: 5 Questions
edited by Vincent F. Hendricks & Hannes Leitgeb
January 2008

Philosophy of Computing and Information: 5 Questions
edited by Luciano Floridi
Sepetmber 2008

Philosophy of the Social Sciences: 5 Questions
edited by Diego Ríos & Christoph Schmidt-Petri
September 2008

Epistemology: 5 Questions
edited by Vincent F. Hendricks & Duncan Pritchard
September 2008

Probability and Statistics: 5 Questions
edited by Alan Hájek & Vincent F. Hendricks
November 2008

See all published and forthcoming books in the 5 Questions series at
www.vince-inc.com/automatic.html

COMPLEXITY
5 QUESTIONS

edited by

Carlos Gershenson

Automatic Press ♦ ⊻⊢P

Automatic Press ◆ $\frac{V}{I}$P

Information on this title: www.vince-inc.com/automatic.html

© Automatic Press / VIP 2008

First published 2008

Printed in the United States of America
and the United Kingdom

ISBN-10   87-92130-13-5   paperback
ISBN-13   978-87-92130-13-6   paperback

Typeset in LaTeX2$_\varepsilon$
Cover illustration by Carlos Gershenson
Graphic design by Vincent F. Hendricks

Contents

Preface

Complexity: 5 Questions

It is difficult to define the interdisciplinary field of complexity. This is noticeable in the often contradictory views compiled in this book. The intention of the volume is not to reach a consensus, but to present the different opinions that dominate the field. Contributors have different backgrounds, including physics, economics, engineering, philosophy, computer science, sociology, biology, mathematics, and chemistry. I do not see this as a problem, in the sense that one view should be right and all the others wrong. On the contrary, I see the variety of opinions as a benefit, since we have different descriptions from different perspectives on the same concepts. From all of these perspectives we can learn more than we could from a single one.

The chapters are organized alphabetically, but can be read in any order. Every contribution is different and interesting in its own way, and there is no definite set of answers for the five questions.

I will not attempt to describe complexity thoroughly here. Instead, I invite the reader to discover many of its facets in the following chapters. I will limit myself to a brief introduction.

Etymologically, complexity comes from the Latin *plexus*, which means interwoven. A complex system is one in which elements interact and affect each other so that it is difficult to separate the behavior of individual elements. Examples are a cell composed of interacting molecules, a brain composed of interacting neurons, and a market composed of interacting merchants. More examples are an ant colony, the Internet, a city, an ecosystem, traffic, weather, and crowds. In each of these systems, the state of an element depends partly on the states of other elements, and affects them in turn. This makes it difficult to study complex systems with traditional linear and reductionistic approaches.

One of the main features—and problems—of complexity is that it can be found almost everywhere. This makes its study and understanding very appealing, since complexity concepts can in principle be applied to different fields. However, its generality makes it difficult to define, and it is sometimes too distant from practical applications, which may lead to its dismissal by pragmatic researchers.

One of the most debated aspects of complexity in this book is its status as a science. Some people agree that it is already a science. Some believe that it is too early to call it a science, but that with time it will mature and become a rightful science on its own. Others think that complexity will never become a science, due to its vague nature. In any case, the study of complexity has scientific aspects to it, but also has been applied as a method to solve problems. Moreover, concepts developed within complexity have been absorbed by well-established fields. Still, these are not referred as "complex biology" or "complex physics". It might be the case that all fields will soon absorb complexity concepts. Will they use the term 'complexity' to refer to them? It does not matter, that is just a convention.

————————————— ♦ —————————————

Carlos Gershenson
Cambridge, MA, USA
September 2008

Acknowledgements

I thank the contributors to this volume for their time, enthusiasm, and ideas. I trust that their work will be an inspiration for future generations of complexity researchers.

John Holland, Marko Rodriguez, Clément Vidal, and Mark Woolsey gave me useful advice.

I thank John Symons and Vincent Hendricks, the series editors, for inviting me to edit this book. They have been very supportive and I owe them for this wonderful experience.

I dedicate this book to Nadya and Ana, who fill my days with hope and grace.

Carlos Gershenson
Cambridge, MA, USA
September 2008

1

Peter M. Allen

Professor

Complex Systems Management Centre, School of Management,
Cranfield University, UK

---

1. Why did you begin working with complex systems?

My PhD was, by chance, about trying to explain macroscopic properties of dense fluids such as viscosity and thermal conductivity on the
basis of the details of their interacting atoms, thus launching myself,
without realizing it, into the statistical mechanics of non-equilibrium
open systems. This introduced me to the work of Ilya Prigogine and
his group at Brussels University and after my PhD in 1969, I went to
Brussels to continue my research as part of Prigogine's group. These
were exciting years and during the next decades I participated in the
research going on there, and at Prigogine's other Centre at the University of Texas at Austin. The discovery and modelling of spontaneous
pattern formation in chemical systems, and generally of Dissipative
Structures, was a vital advance, and I soon became interested in ex-
amining the importance of these ideas in ecological and human systems, since chemical kinetic equations are simply a "special case" of
population dynamics.

This led me to consider the relationship of this work to Catastrophe Theory (René Thom) and then to Non-Linear Dynamical systems
and bifurcation and theories of attractors. Complexity emerged as the
theory underlying non-linear dynamics in that these equations were
always averages of fluctuating variables, and the trajectory of the
system therefore resulted from the interplay between the trajectory
of the system, the attractor landscape in which it moved, and the
fluctuations that could potential drive the system from one attractor
basin to another. My own work looked at the emergence and evolution
of spatial patterns in ecological and human systems leads to models
of urban and regional systems. Beyond this, I became interested in

the behaviour of the system subjected to fluctuations not only in the values of existing variables, but also from the heterogeneous, micro-diversity of the individual elements or agents, focusing on the issue of a system's structural stability, and developing, from the late seventies, models of qualitative evolutionary change. For me, complex systems science differs from systems science in that it includes the qualitative, structural evolution of systems, how they became what they are, and how they may evolve into the future. This has led me into models concerning the resilience, evolution and "management" of: natural resource systems (fisheries), economic markets, organizations and integrated spatial models (employment, households, transport and environment) of urban and regional systems.

2. How would you define complexity?

I think that it is what characterizes all evolved, open systems, where the structure and organization has emerged over time through processes of self-organization. It therefore applies to all natural, multi-level systems (molecules, cells, organs, organisms, groups, populations, ecosystems, etc.). Many artefacts however may well not be complex in themselves, being perhaps simply mechanical, but when we include the designer and the system which drives the production of artefacts then we once again arrive at complexity. There is an open ended socially constructed process which is path dependent and both driven and limited by culture and history. Complexity emerges and evolves as a result of heterogeneity which is created by underdetermined processes, and undergoes differential dynamics in which some elements are amplified and some are suppressed. This continuous process leads therefore to successive dynamical systems, which call structural attractors (sets of phenomena that together have ways of getting resources from the environment. This process of qualitative, structural change is that of qualitative evolution and which can be temporarily stable, until some new micro-heterogeneity is amplified, changing the boundary of the previous system, and giving rise to emergent properties and features. In a 1987 paper (Allen and McGlade, Foundations of Physics)) this was defined as evolutionary drive—the evolutionary selection of the power to evolve.

3. What is your favourite aspect / concept of complexity?

I think that it throws a remarkable new light on the human situation, one quite different from that of traditional science. It confirms that the future does not exist, and that therefore we are part of its creation. This means that although sometimes the system we are in may seem to offer unpleasant outcomes, we can perhaps hope for, or create, an instability with some new dimensions and aspects which can open new, more exciting vistas. This is fundamentally optimistic, and rests paradoxically on the limits of knowledge, which allow us the freedom to explore and create new systems. It tells us that our interpretive frameworks will keep on evolving, and therefore that there will be different understandings of reality—that communication and exchange can therefore be valuable, but that it will always be impossible to explain our experiences on the basis of any simple principle, other than that of recognizing that the world is bound to be messy. Sense-making may be more of a creative act than we like to think, rather than the revelation of some underlying intelligence.

4. In your opinion, what is the most problematic aspect / concept of complexity?

The most problematic aspect is that this acceptance of an evolving world in which there are severe limits to real knowledge and prediction runs contrary to the world of simple "bottom lines" and heroic capitalism in which we live. In today's world it is often assumed that disproportionate success arises from internal, intuitive brilliance, whereas in reality complexity tells us that it will often arise from luck or privilege. The idea of adopting an open, learning, humble, reflective and exploratory approach is not fashionable, as it does not make a media splash. Complexity tells us that in reality, everything is an experiment—and we need to consider and question what is happening and why, if we are not to be surprised and destroyed as in reality so many companies and organizations are.

5. How do you see the future of complexity? (including obstacles, dangers, promises, and relations with other areas)

I see the future of complexity as becoming more and more that of the real world. At last policy and decision support is recognizing the need for interdisciplinary approaches rather than separate expertises, and this multi-perspective view offers considerable improvement compared to the childish mono-dimensionality of economics. The recognition of the irreducibility of the world to any single dimension of comparison will be a great advance. Clearly, there are also dangers that complexity may be dismissed as a mere "fashion" in the social sciences where there is much over-hype and superficial, empty rhetoric. In fact the science of complexity unites both the "hard" natural sciences and the "really difficult" ones involving biological and social phenomena. In the end complexity indicates how the universe could give rise to biological and human phenomena, without necessarily supposing a supernatural entity. Indeed it shows us that the creativity of complexity is quite natural. At some point therefore, complexity could offer a point of reconciliation for humanity—though not for some time.

2

Philip W. Anderson

Professor Emeritus

Physics Deptartment, Princeton University, USA

The Nobel Prize in Physics 1977

---

1. Why did you begin working with complex systems?

In a sense, I have been into complexity all my professional life, since my thesis was about explaining measurements on a macroscopic bag of gas in terms of the quantum mechanics of the separate atoms; and I found this aspect of physics so fascinating that i never looked back. What I was doing was was an early exercise in what was later to be called "many-body physics" and the philosophical idea is that the quantum theory is not just for atoms but should be able to give you understanding of matter at any scale.

I came across, in the '50's, two phenomena which went deeper into the roots of complexity: the first was broken symmetry in antiferro-magnetism and superconductivity, which is the best understood and most primitive example of emergence in physics—and it was this that led me eventually to write in 1967 "More is Different", my entree into the complexity world. The second was localization of quantum-mechanical waves, which launched me into the world of disorder, ran-domness and the like—it showed that qualitative change could result from dirt and disorder (and incidentally won me a Nobel Prize.)

Still in pure physics, in 69–77 I got involved in spin glass, which turned out to be a major influence in all kinds of complexity problems—computational complexity problems, algorithms for complex optimization, neural network properties, coding, you name it. A direct line runs from this work to this year's [2006, ed.] Les Houches summer school on complexity chaired by Marc Mezard.

There were several separate influences which drew me into the world of complexity beyond physics. One was the little group of Gene Yates,

Ibby Iberall and friends (Mandell, Shaw...) who looked me up as a consequence of "More is Different" and involved me in a couple of meetings in the late 70's, especially a Dubrovnik meeting on self-organization where I met people like Abrahams, Stent, Orgel, Tudor Finch, Nicolis, Harold Morowitz, from whom I learned a lot about biology. There is a book, to which I contributed the opinion that dissipative structures are not the secret of life. Second was a couple of physicist's meetings, one hosted by Werner Erhard of all people, where I met Stu Kauffman and Norm Packard, and another organized by John Hopfield at Aspen with the Thinking Machines crowd and others (Hopfield was always an influence). But this all came together when David Pines involved me in the early organization of SFI, and I remember having the feeling that I had been solving complex problems all my life, and certainly belonged there.

2. How would you define complexity?

I think it's a mistake to try to define complexity. I'm happy to define emergence, as the appearance as the scale is increased of properties unrelated to those of the substrate. I think one knows complexity when one sees it.

But that is a cop-out—let's give it a try. How about—the search for generalizable features in the behavior of large or complex aggregates of simpler systems. There is a hope that these features may be described mathematically (like chaos or fractals) and/or that they can be related to the properties of simpler subsystems.

3. What is your favourite aspect / concept of complexity?

I don't like to pick a favorite, nor see any basis for doing so. My first paper on complexity proper was on the origin of life, and I still think that one of the major intellectual puzzles of our time. I do think it not hopeless. I agree with Stu Kauffman (and, actually with ideas of Iberall) that the key is work and energy—life, even reproduction, requires doing work.

The spin glass still intrigues me. For one thing, it is a real, calculable example of self-organized criticality. The principle of SOC has had a bad press, but it is real, though it is not "how the world works".

The really big problems are consciousness and free will and anyone who isn't intrigued by these isn't alive intellectually.

4. In your opinion, what is the most problematic aspect / concept of complexity?

Whether it can ever deal with economics except in the trivial if rewarding field of finance. Of course, there is always the temptation to believe that your computer can do your thinking for you.

5. How do you see the future of complexity? (including obstacles, dangers, promises, and relations with other areas)

I see the future of science and the future of complexity as converging on each other. More and more science is learning interdisciplinarity and there will be fewer and fewer simple answers that are just chemistry or biology or physics.

3

W. Brian Arthur

External Professor

Santa Fe Institute, USA

Visiting Researcher

Intelligent Systems Lab, PARC, USA

Schumpeter Prize in Economics 1990

---

1. Why did you begin working with complex systems?

I am sure that like a lot of other people, I did not really begin working with complex systems. I was working on particular problems and I got attracted into reading things that turned out to have to do with complex systems. Quite a while ago, in 1979, I read several books that influenced me. One of them was by Hermann Haken. I was very influenced by that book. I was reading very widely, and I started to read in molecular biology. I also read Horace Freeland Judson's "The Eighth Day of Creation", which is about molecular biology and the history of decoding DNA, its transcription structure, and how proteins are put together. Somewhere around June, 1979, I read an essay by Ilya Prigogine on self-organizing systems, and that had an enormous influence on me as well. I was either reading books that had to do with how biology works, or reading books that had to do with positive feedback, and Prigogine was talking about the chemistry of self-catalyzing systems. So it began to become very clear to me how I could apply this to economics.

2. How would you define complexity?

I do not think of complexity as having a definition. Seth Lloyd compiled 45 definitions of complexity. I see complexity not so much as something that we define, but more as a movement. It is really a movement to look at how multiple elements reacting to the patterns that those elements together create, as opposed to the reductionist approach, which looks at the things from the top down in finer and finer detail. All of the systems that I know that seem to be complex have a vague definition. They could be stars in a galaxy, cells in an immune system, consumers or investors in an economic market, ions in a spin glass. But they are always multiple elements reacting to the patterns that those elements create. Some systems that answer that description would give you an interesting result, or they might just tend to some equilibrium and stay there. Other systems might pick one quasi-stable state out of a huge multiplicity of possible ones. And other systems might go on and generate new patterns. Then others might never stop generating new patterns, in actual perpetual novelty.

I see complexity as a movement that came along in the sciences, sometime around the 1970's or 1980's. There are many precursors to that movement. Poincaré and others, even Darwin, you could claim, thought about how systems of many elements unfolded and how patterns were created by single elements. In the 1980's we got desktop computers and we could easily simulate those systems. Instead of simplifying these systems by trying to write a single equation for them, we could allow the computer to model them and see on the screen how these unfolded over time. That, for me, was when complexity started to take off. Not so much as a theory, but just as an approach to do science. It is complementary and in part in competition with the reductionist view, where you look at species, and then at organisms, and then cells, and then structures within the cell, in finer and finer detail. Complexity looks at things from a very different direction. That definition as a movement tells me that it is not going to go away in ten years or so. It will probably be with us for the next several centuries, as people look more and more systems unfolding and evolving over time.

3. What is your favourite aspect / concept of complexity?

I love looking at systems that are evolving over time. I have seen many types of nonlinear stochastic processes in two or three dimensions and economic systems, such as the stock market. I just love looking at the computer screen, watching these things evolve over time. And in particular I am fascinated by systems that do not tend to any steady state, that are always showing perpetual novelty. I do not know why, but it strikes me that there is a truth in nature that our mind has not been able to capture very well, and that is all the interesting systems we know of, the set of species, the ecosystems, they do not tend in the very long run to single equilibrium or steady state. New species are generated and novelty never stops.

There is a lovely paper by Lindgren[1] looking at an iterated prisoner's dilemma scheme where the strategies are the elements. There are many strategies competing against each other, but there is no steady state. New strategies are discovered perpetually and I find that fascinating.

4. In your opinion, what is the most problematic aspect / concept of complexity?

For me, the most problematic aspect of complexity is thinking that there is a theory of complexity. I do not believe that there is any theory of complexity. I believe that complexity is an approach to do science. There is no theory of reduction in science. There are no theorems that would define that. And I do not think that there are theorems or a particular collection of explanations that would constitute a theory of complexity. Stuart Kauffman might disagree with me, but I have never believed that there should be or will be a theory of complexity, and I think that it is problematic if people start looking for one. Still, I may be proved wrong that there may be very deep laws that remain to be discovered.

The other problematic aspect is that we tend to expect results in a very short time. I remember when the Santa Fe Institute started in the late 1980's, within five years, journalists were asking us what results had we found "after all, it has been five years, almost approaching ten years, it is about time you guys settled all this". But I do not think it is like that at all. I think it is going to take decades or centuries for science to see what complex systems really are all about. Progress will be slow, but that echoes progress in the other approach of science, reductionism. New ideas and new insights were built as new instruments came into being like the microscope or the telescope. Theories followed measurements, and this took decades, not years. So I think

---

[1]Lindgren, K. (1991). Evolutionary Phenomena in Simple Dynamics, In Langton, C.G., C. Taylor, J. D. Farmer, & S. Rasmussen (Eds.) *Artificial Life II*, pp. 295–312. Addison-Wesley.

that the deeper insights will take decades more to come. So far, we only got a taste of what is coming.

5. How do you see the future of complexity? (including obstacles, dangers, promises, and relations with other areas)

I think there is a danger in expecting too much. As I said, we should not expect just to snap your fingers and there will be theories of complexity that we can then apply to economics, biology or neuroscience. I do not think that it is that way at all.

Darwin's idea[2] of variation and selection comprised in the mechanism for evolution has been applied in many fields in the hundred and fifty years or so since Darwin. It has been applied to behavioral biology, in economics, in linguistics, in areas of philosophy, in the evolution of human behavior, and so on. People find that there is a wealth of insight that normally would not have been arrived at with Darwin's basic idea.

I think that it is much the same with complexity. Seeing fields as comprised of elements that are adapting to or reacting to the pattern they together create gives you a very different view of economics, of sociology, of systems within the brain, neural structures, even of certain parts of chemistry. I do not think that the promise is that—as that particular set of insights worked its way through each field in science—it will add to, and sometimes even transform, the field.

In my own field, economics, the complex systems approach is not just an addition to the standard approach. The standard approach says: Take consumers, producers, investors, and business decision makers. If you formulate problems in which they take part, you can identify and analyse an equilibrium. That gives you equilibrium economics. If you start to look at those same consumers, producers, investors, and decision makers reacting to a situation they together create and making new decisions and new reactions and new business plans, you are in a situation that may never arrive at any equilibrium. So, complexity economics means, in general, doing economics in non equilibrium or out of equilibrium. So complexity is not just a minor addition to economics. It is actually the flip side of standard economics. We have known two hundred years of very good work with the economy as an equilibrium system.

---

[2]See Dennett, D. C. (1995). *Darwin's Dangerous Idea: Evolution and the Meanings of Life.* Simon & Schuster

The complex systems approach is showing us how to think of the economy in non equilibrium. And just as nonlinear anything is much more complicated and interesting that linear, non equilibrium is much more interesting and much more general than equilibrium. Equilibrium systems are a very special case of non equilibrium systems.

4

Yaneer Bar-Yam

President
New England Complex Systems Institute

---

1. Why did you begin working with complex systems?

The real issue is what are the questions we are interested in answering. After I finished my PhD in physics in 1984, I didn't feel constrained by the questions that were being asked by others in physics. I wanted to understand many things about the world around—human thought, society (particularly civilization)—as well as to consider some of the problems people I knew were concerned about—polymer dynamics and protein folding—and then other topics in biology—developmental biology, evolution, and so on. When I explored these questions I found that dramatic progress could be made. At MIT, I had been exposed to ideas about phase transitions, cellular automata, spin glasses, and system dynamics. These provided some of the tools I used. When I needed tools I didn't have, I had to construct them or was able to find them in other fields. However, when I tried to explain my ideas and results to others, it was often difficult to find a way for colleagues to relate to what I was explaining. The connection between the various things I was working on was also not apparent. At one point someone said to me: "You are working on so many different things." I responded "No I'm not." Over time the underlying themes of interdependence, collective behavior, patterns, self-organization, complexity, and emergence—themes that violate the way traditional science thinks about the world—became clear. The tools that I accumulated became a powerful and general approach for thinking about complex systems.

Eventually, I decided to write a textbook that would explain what I was interested in—Dynamics of Complex Systems, published in 1997—and taught a course from my notes for the book starting in academic year 1993/4. To my surprise the course attracted not only students but many faculty, even from other institutions. The course format gave me the time to introduce key ideas that could be used to make progress in the understanding of complex systems. After the basic ideas were present, I could explain the questions that I was exploring, the new results, and the underlying themes and tools. While there was great interest in my classes at the local level, it was still hard to have the ideas accepted more widely and to communicate results through journals.

Over time, I became aware that there were others who were engaged with these or similar ideas. This was not clear from the beginning as there was no framework for me to discover their work. The historical roots of systems ideas became more apparent, as they exist in decades (or longer) of contribution to systems thinking across disciplines throughout the sciences and in engineering. These ideas were often not found in standard educational programs. By and large the more recent developments in complex systems ideas at that time didn't have a positive name in the academic community. Nevertheless, they were attracting popular interest through the books of Gleick (Chaos), Waldrop (Complexity) and Lewin (Complexity). I expect that this may have annoyed some traditional scientists, but more commonly they were probably not even aware of these ideas and surely the power of the new ideas had not been sufficiently demonstrated to them. We had just started, and while we were excited about what we could do, the responsibility for demonstration was ours. It was clear that a more formal framework both for interactions between scientists and for the research itself was needed.

Building on the interactions with the faculty attending my classes we decided to launch the New England Complex Systems Institute (NECSI) as a center to advance research and education. At that time, interdisciplinary (or better yet transdisciplinary) work was a radical move and largely impossible in the existing academic institutions. Young faculty were, and in some places still are, discouraged from deviating from the narrow path, as tenure would be impossible if there wasn't a clear and focused domain within which they could be recognized as experts. It is important today to realize that the pioneers in this field put themselves at great risk. This tends to be forgotten as the acceptance becomes wider and many people are happy to be associated with interdisciplinary and complex systems research. The pioneers protected themselves as best they could, but there were definitely injuries and scars that remain.

The first task of NECSI was to organize a conference, the International Conference on Complex Systems, to declare the importance of the new field and give a venue for people to meet and discuss what we were interested in. I invited many great researchers with broad vision to present at the conference. Still, most of them did not understand the unifying concepts of complex systems. At the first conference many of them stood up and started their lecture with the exact words: "And now for something completely different". They didn't see the connection to the talk before theirs. By the second conference this was said by fewer people and by the third or fourth one we stopped hearing it.

Traditional science is defined by the use of calculus and statistics and the limitations that were inherent in paper and pencil studies. When new ideas were developed in the 1970s and computers became available, it took some time before people became aware that the limitations that were imposed by previous approaches could be discarded. Many scientists, though surely not all, did not recognize how the limitations impact the way they think about the world. The assumptions of calculus (smoothness) and statistics (independence) are violated by developments in statistical physics of phase transitions that were made in the 1970s. The assumptions of calculus are also violated by fractals and chaos, developed also in the 1970s. Other ideas, such as pattern formation, were introduced even earlier. Still, it takes time for people to recognize how deeply these concepts change the way we think about the world.

2. How would you define complexity?

Complex systems and complexity refer to the existence of system behaviors that cannot be described from the behavior of parts, and must be described through understanding their interdependence. Interdependence couples the behaviors of the parts. Complexity also has a quantitative definition—a definition that enables us to quantify how complex something is. Various researchers have worked on a definition of complexity in this sense. Perhaps the main difficulty is that the most straightforward definition seems to have insurmountable problems. This is the idea that complexity is the amount of information necessary to describe a system. The problems range from subjectivity to the fact that we need more information to describe disorder and disorder doesn't seem to be what we mean by complex.

Rather than creating a different definition, I realized that the key issue is that complexity depends on the scale of observation. By considering the complexity as a function of the scale of observation, and not trying to consider a single scale, we can not only define complexity in a consistent and meaningful way, but we can also learn a tremendous amount about complex systems. The Complexity Profile, the complexity as a function of scale, is a conceptual and quantitative tool that I have used to address many questions about real world complex systems. Often, when people ask me to help them understand problems they are dealing with—whether the US military, the World Bank, healthcare executives, people building highly complex engineering projects or policy makers—the complexity profile helps to explain the origins of the problem in the complexity of tasks that are to be performed, and thus also how to better approach solving it.

Thus, my quantitative definition of complexity at a particular scale is the amount of information needed to describe the system at that scale of observation. The amount of information is quantified through Shannon information theory or perhaps another formalization. This definition makes the complexity a function of scale—the Complexity Profile. It is a concept that is not contained in any other framework I have seen and is remarkably powerful in its explanatory capabilities.

3. What is your favourite aspect / concept of complexity?

The study of complex systems is remarkable in the ongoing nature of discovery. New and deep insights occur frequently. This is quite unlike traditional fields of science where tremendous efforts are needed to make progress. For almost 10 years I have been teaching a weeklong course, now part of our summer and winter schools, in which I focus on the basic ideas that I have found to underlie my research in complex systems science. Almost any one hour lecture in this course brings us to the frontier of what is known and provides a direction and opening for new contributions. Consider that for traditional science only after years of specialized study of a discipline is it possible to get to the point where a contribution can be made to important questions. By contrast, in complex systems, we can readily identify many questions that are of great interest in an hour and the possibility of contribution becomes apparent—though surely understanding the many concepts and tools involves much more extensive engagement.

4. In your opinion, what is the most problematic aspect / concept of complexity?

I think the most problematic aspect of complexity is that our lives are entangled with it. We are experiencing an explosion of complexity through changes in technology and society. Because of the increasing complexity, the social structures that were effective only a few years ago are no longer able to perform their tasks. We have seen over and over again how organizations, including healthcare, education, military, and government in general, must change to deal effectively with the complexity of their tasks.

The rise of complex systems science seems to be intimately tied to the rise of the complexity around us. We need it in order to make sense of the world. Still, most people are trained to think in traditional terms. In a traditional reductionist perspective, social systems are effective if a good leader is in charge—most forms of government, from monarchies to dictatorships to democracy, are about the way to select the right person to be in charge. If something goes wrong, we need to make someone responsible, put the right person in charge to fix it. Today, more often than not, having someone in charge is the problem. To make almost any system work effectively, we need to distribute responsibility and decision making in dramatically different ways and enable the necessary communication and coordination. Unfortunately, many people are dying (in military conflict, from medical errors, in natural disasters like Katerina) and there is much suffering in the world because this lesson is hard for people to learn.

My confidence that complex systems science can be an important social policy tool has grown through the experience I have had in addressing real world problems. Some of these experiences I describe in the book Making Things Work. I think the promise of our science underlies the early popular interest in the field. People sensed that there was something about what we were engaged in that connected with their day to day experience in the real world. Chaos and self-organization, are all around us. The business management literature adopted complex systems ideas very early, already in the 1980s and 90s, recognizing the limitations of centralized control and the need for alternatives. Both the science and its application to management have made much progress since then, and more needs to be done. Still, it is safe to say that the early interest existed because of a deep connection to the new ideas. Today complex systems science is surely in a position to help with real world problems.

The financial crisis which is unfolding as I write this in October of 2008 is just the latest example, perhaps a telling one, of how traditional ideas and traditional social systems are not able to deal with complexity. The result is an ongoing transformation of society that is dramatic and serves as the real time laboratory for understanding complex systems in action.

5. How do you see the future of complexity? (including obstacles, dangers, promises, and relations with other areas)

I once ambitiously believed that complex systems science would become a new field of science. Today I believe that it will become at least as rich a domain of inquiry as all of the traditional fields of science and more. Moreover, I believe that traditional science will be transformed by the existence of complex systems science. It is traditional science that was a limited (low dimensional) view of the world ignoring much that is important. Complex systems science adds new dimensions so we can see what is around us and powerful tools to think about them. Complex systems science sheds new light for our understanding of the world.

The biggest current danger to the field is that it will be hijacked by people who don't understand the essence of the field. Many are adopting the terminology without understanding what complex systems is really about. Systems biology, systems engineering and other systems related fields are often (but not always) just using the words but continuing a reductionistic approach. Moreover, there are many who do not understand how to properly use the tools that are available, like computer simulation. This leads to support for projects that cannot succeed and can give our field a bad name at a time when we are developing important credibility in science, and with people confronting real world problems who need our help.

Ultimately, there is no way to go but to develop complex systems science. It is the science that can address the questions about the world around us—because the world is complex. We will need all of the rich tools and concepts we can develop to answer important questions and address challenges we are and will increasingly be facing. Moreover, the ideas of complex systems science can be related to everything that we care about, every aspect of human creativity, including the humanities. Complex systems is about understanding what we know and how we know it, and most importantly, the world around us.

5

Eric Bonabeau

Chief Scientist

Icosystem Corporation, USA and France

---

1. Why did you begin working with complex systems?

How exciting can it be working with simple systems? More seriously, I have been in it since 1990. As a research engineer in the R&D unit of France Telecom working on telecommunications networks, I had to look for ways to understand, predict and hopefully control complex, distributed systems that operate in dynamic environments. In telecommunications networks, traffic and topology change all the time, for example links or nodes break down, come back up, or new nodes and links may be added when the network is upgraded, and you always need to transmit messages regardless of the network conditions. I began some scouting work. After stumbling on work by Jean-Louis Deneubourg of the Free University of Brussels on social insects, I became convinced that the "swarm" metaphor could be fruitful for managing large communications networks. As a result I started work on swarm intelligence at France Telecom, developing algorithms relevant to communications networks. I then became interested in tools with even broader applicability to telecommunications problems, including agent-based modeling, evolutionary computation, complex landscapes. I also started to develop a taste for the study of social insects, who were at the heart of the metaphors I was using. My search for like-minded people led me to the Santa Fe Institute, where I spent 3 years as the Interval Research Fellow after leaving France Telecom in 1996. I must say that, once you start working in this field, it becomes quite addictive. I mean, how can you not be excited by a discipline that is interested in genes, organisms, groups, societies, networks, economies, institutions, chaos, robustness, evolution, and more?

2. How would you define complexity?

While I have no single definition of complexity and how to measure it (I think it is problem- and observer-dependent), I have a rather traditional and boring, if physics-minded, definition of a complex system: it is a system comprising multiple, sometimes many, constituent units that interact in such a way that the system's aggregate behavior cannot be inferred from the behavior of one constituent unit in isolation. In other words, it is a system you need to study holistically if you want to understand how it works. The definition may be boring, but its implications are anything but: perhaps the most intriguing one is that it applies to many, many different systems from different disciplines and it suggests that a method borrowed from one discipline can be used to understand a completely different system in a completely different discipline operating at completely different scales. For example, the tools of statistical mechanics, originally developed to describe the collective properties of physical particles, can generate insights into the collective behavior of social insects or the crowd behavior of humans. Obviously, one should exert caution when importing from and exporting to other disciplines, as the similarity is at the level of the tools and should not be assumed to have any validity beyond that. For example, an ant is not a particle, although it can, for the purpose of understanding the collective dynamics of a colony, be modeled as a particle-like entity; similarly, a human being is neither an ant nor a particle, but it can be useful to use that level of abstraction for the purpose of understanding the collective behavior of a crowd—indeed, the behavioral repertoire of humans can sometimes be constrained to just a few options. So for me, finding inspiration from disciplines other than yours is a hallmark of complexity science.

3. What is your favourite aspect / concept of complexity?

What I find fascinating about complexity science is concept and tool transfers between fields. This is different from saying that complexity scientists are on a quest for universal laws—at least I have never considered this to be a realistic objective. On the contrary, transfer of concepts and tools occurs on a case-by-case basis.

   Consider my own trajectory. First, I have used my knowledge of many-body systems from statistical physics to model social insect societies. This knowledge transfer is neither straightforward nor trivial. As a statistical physicist, you need to understand the detailed mechanisms and constraints of how social insects operate, and you need to familiarize yourself with the body of experimental work, otherwise you are just building metaphors. Second, I used my knowledge of collective behavior in social insects, formalized with the help of models imported from statistical physics, to design algorithmic systems inspired by social insects. Taking insights from social insects and applying them to engineering problems is neither straightforward nor trivial. You need to understand the fundamental principles that enable social insect societies to function as coherent wholes, and then you need to understand the specifics and constraints of the engineering problem you are addressing. The difference here, which to me makes the concept transfer toward engineering perhaps more exciting, is that you are not trying to explain experimental data: as an engineer you are free to invent anything you want, provided it is consistent with the laws of physics.

4. In your opinion, what is the most problematic aspect / concept of complexity?

For complexity as a field of scientific endeavor, the challenges are (1) to continue to foster innovative, cross-disciplinary scientific thinking, which requires considerable energy, and (2) keep clear of claims of universality or universal laws. The ability (or the permission) it gives scientists to use insights from other disciplines is of great enough value to justify the existence of a complex systems community.

But because it is easy to make outrageous claims when importing concepts from other disciplines, it is important that scientists toying with complexity be empirically minded. That's not just making sure the model fits the data, as many a theoretical physicist seems to believe. It also means making sure that the model is consistent with existing knowledge of mechanisms, looking for alternative models, discussing models with specialists, etc.

In other words, if complexity is to succeed in science, it has to behave like a scientific discipline, and therein lies the rub: can it remain fun and exciting as it becomes more mainstream?

But for complexity as an engineering toolset, I don't think that is going to be an issue. There is an unlimited supply of ideas from physical, chemical and biological systems that we can use for engineering purposes. The potential for concept transfer therefore is infinite, full of future surprises and tremendously exciting for any curious mind.

5. How do you see the future of complexity? (including obstacles, dangers, promises, and relations with other areas)

The future of complexity science is already here. There are still walls that need to be taken down, but there is already a new generation of scientists in place for whom thinking across disciplines is just part of the daily routine of being a scientist. That is the tremendous achievement of just 20 years of complexity science.

As to the practical applications of complexity thinking, what I call complexity engineering, we can already get a glimpse of the future, but in many subtle and sometimes hidden forms. For example, we are already seeing a broad familiarity with some of the fundamental concepts of complexity science in the general public in the form of social networking, peer-to-peer systems, wikipedia, web 2.0, and other decentralized systems. Web 2.0 in particular is a great playground for complexity engineers who recognize that their toolkit can help them make sense of the emergent phenomena that take place when millions of individuals start interacting. Google is probably the most successful "complexity company" on earth, although most people and financial commentators would probably drop "complexity": Google understood early how to exploit the emergent knowledge base embedded in the interactions between web pages—a form of information social network of web pages created by users linking their own pages to others'. When I look at the techniques and algorithms used by Google's search engine, they remind me of social network techniques developed in the 50's and 60's and transferred to other disciplines by complexity scientists in the 90's before Google found the "killer app" for them. I am quite certain that other complexity killer apps will continue to emerge along the same lines.

My own path has taken me into a different direction than my swarming past might have suggested. I am very much a complexity engineer, but my metaphor of choice now is not so much insects as evolution, and in particular directed evolution. Evolutionary algorithms are a fantastic success story of how a natural phenomenon (well, not everyone agrees that it is a real natural phenomenon but I am a believer) can be exploited for engineering purposes. One approach I find particularly promising was introduced by Richard Dawkins 20 years ago: interactive evolution, which is the in silico version of directed evolution.

There is an intriguing parallel between biological evolution and decision making: search and evaluation are similar to variation and selection. Nature thus provides us with a powerful metaphor for decision making. Computational techniques known as artificial evolution or evolutionary computation replicate in silico the way that biological evolution works. Since its introduction thirty years ago, evolutionary computation has proven highly successful at solving a wide range of decision problems. In interactive evolution, variation is performed by a non-human device while options generated by the device are evaluated by a human being. In fact, we humans have been using this technique for hundreds of years, it is known under various names such as breeding, animal husbandry, or directed evolution. To please everyone, I like to call it "intelligent design by means of evolution". To name one famous example, corn was bred about 9000 years ago by Mexican farmers. Teosinte, the plant they started with, is so different from modern corn, that it was originally classified in a different genus. Teosinte is barely edible, while corn is today one of the leading sources of calories for human society. The story of how such a transformation was made possible, by the combination of careful selection by farmers with a genetic structure that enabled dramatic morphological changes, is still being uncovered by ongoing research. Which means that humans have been using a powerful biological engine called variation which they did not understand at all; all they knew was that it worked for producing the requisite amount of variation and they could provide selective pressure. Imagine now the same process with the biological engine responsible for variation being replaced by a computing engine. The result is called interactive evolution or IE. IE is very useful when the space of potential solutions, designs or strategic options is large AND the goodness of a solution is difficult to formalize. For example, there are many situations where the decision maker doesn't know ahead of time what the solution looks like—"I know it when I see it" kinds of situations. Starting with a more or less randomly generated population of solutions, the evolutionary technique will search the space of solutions by picking the fittest individuals as defined by the user, will mutate them and breed them, and the offspring will again be evaluated by the user, etc, until solutions emerge that satisfy the user. There have been a number of business applications over the last few years and the trend is accelerating. Honda is helping its designers explore the space of car designs using interactive evolution. The problem with car design is that it is highly constrained: a designer has to satisfy hundreds of technological constraints simultaneously (such as wheelbase length, windshield angle, and size of engine compartment) while at the same time remaining creative. In other words, automobile designers must balance aesthetic considerations with technical specifications, an often frustrating juggling act resulting in a lengthy trial-and-error design process. The tool enables the designer to engage in a guided exploration of the design space: it is first presented with a number of initially random designs; the designer picks the ones that come the closest to what he is looking for—they are the fittest individuals; artificial evolution takes the fittest designs, mutates them and breeds them to create a new "virtual generation" of designs, which the designer evaluates again. The results are spectacular: in just a few iterations, a car designer can create any design he wants consistent with the constraints. Designers can create and compare a vast number of designs in a short time, greatly streamlining and accelerating the design process. Icosystem has applied IE to aircraft design, control system design, intrusion detection on computer networks, postal route optimization, drug discovery, exploratory data mining and model calibration. Companies like Procter and Gamble or Pepsi-Cola North America have harnessed the power of evolution combined with the collective brainpower of their customers with the help of Cambridge, MA-based Affinova to come up with new packaging or product designs. It works very much in the same way as the Honda tool but it is consumers who pick the designs they like as opposed to professional designers. Companies are able to directly translate their customers' potentially non-verbal tastes and preferences into new products or designs that will please specific segments.

Why do I think IE is such a promising approach? Because it empowers us to explore and invent. IE helps you navigate the design space. You can never be sure you're exploring the space in an exhaustive manner. But you're navigating it in a way that is a lot smarter than a random walk and a lot more empowering than being forced into a solution. That's my complexity killer app.

6

Paul Cilliers

Professor of Philosophy
University of Stellenbosch, South Africa

1. Why did you begin working with complex systems?

During the middle 80's I was working as a Research Engineer on problems in computer modelling and pattern recognition. I was also studying Philosophy as a hobby. During the day I worked on sound recognition using neural networks and at night I read philosophy of language (and changed nappies—not my own). During my postgraduate studies in Philosophy my interests began to focus on the structuralist linguistics of Ferdinand de Saussure and the subsequent positions developed in French structuralism and post-structuralism.

In my work on pattern recognition, I became aware of the limitations of the traditional formal, rule-based models of mainstream AI. From a philosophical perspective, this work was then strongly influenced by the "functionalist" argument, developed by (and later strongly amended) by Hilary Putnam. Functionalism claims that the computer is an adequate model for the brain. My work in neural networks seemed to introduce a totally different understanding of "computer" though: a distributed (connectionist) process, not the formal manipulation of rules. When Fodor put up a strong defence of functionalism against connectionism, thus re-enforcing the traditional approaches from analytical philosophy, it became clear to me that alternative paradigms needed to be investigated.

As part of a shift away from "mind" towards "brain" I discovered the neurological model of the brain Freud developed in his early years. I noticed that this model bears a striking structural similarity to Saussure's model of language, and what is more, that both these models can be described in terms of a recurrent neural network. This insight enabled me to develop a new kind of description of the relationship between brain/mind (Freud), language (Saussure) and distributed models, a description quite different from the Chomsky-Turing-functionalist one predominant in AI and analytical philosophy of mind.

These ideas were developed in a PhD in Philosophy done jointly at Stellenbosch and with Mary Hesse at Cambridge. It became clear that this family of distributed models could serve as a general description of complex systems, not just the brain. The next, and perhaps least conventional step, was to realise that Derrida uses both Freud and Saussure in developing the widely influential philosophical strategy known as "deconstruction". The combination of complexity theory and deconstruction proved to be a novel and fruitful one. It allowed me to "deconstruct" some of the more positivist assumptions in complexity theory, and also allowed me to develop a more rigorous understanding of deconstruction than the rather wayward one then fashionable in literary theory. This approach, as described in *Complexity and Postmodernism* (Routledge 1998), also opened up a wider understanding of the implications of complexity theory for the social sciences.

Currently I work with qualitative descriptions of complexity, rather than with quantitative or computational models. I think there is a great need for "understanding" complexity, and for understanding the limits of the models we use in the process. I do not see this a secondary activity. Philosophical reflection is an essential component of all research.

2. How would you define complexity?

There are a number of philosophical problems with the notion of "definition". A definition—describing something in terms of a set of more primitive concepts—is already reductionist in nature. The distributed nature of complex systems resists a description in terms of essential components or principles. There is thus a certain performative tension between the notions "complexity" and "definition". Nevertheless, one should at the same time attempt to do better than to merely say that "complexity is very complex". Consequently I prefer to discuss a number of related characteristics of complexity, rather than to provide a definition.

The most important characteristic of a complex system is probably the fact that it has emergent properties. The notion "emergence" should not be used to evoke anything mysterious or metaphysical. For me it merely refers to the fact that a complex system has properties which result from the non-linear interaction between the components of the system, properties which cannot be reduced to some property of the components themselves. It is the presence of emergence which precludes a general application of the "analytical method", i.e. of "cutting-up" the system into manageable sub-systems. The converse—trying to model complete complex systems with complex models—is unfortunately just as problematic. Now the model will have its own emergent properties, and will be just as difficult to understand as the original system. We also have no objective way of matching emergence in the model with emergence in the system.

As can be seen from this discussion, the characteristics of complexity—non-linearity, emergence, distributed feedback, non-equilibrium, etc.—do not really help us to pin complexity down. Rather, it shows us why it so difficult to do exactly that. Many of the algorithmic understandings of complexity is not mindful enough of this issue and consequently provide definitions of complexity which claim more than they can deliver. There are many useful definitions of *aspects* of complexity, for example of the "incompressibility" of complexity, but these do not capture "complexity" in any complete or general sense.

Another problematic aspect of our descriptions of complexity is the relationship between epistemological complexity and ontological complexity. To what extent is the complexity of a system an effect of our *description* of the system and to what extent is it a characteristic of the system itself? The problem becomes clear if one thinks of systems which may appear complex initially, complete with emergent properties, but once we get to know the system the emergence evaporates and we understand it quite comprehensively—think of someone's first encounter with a complicated piece of technology. The difficult question to face here is to ask whether there are systems which are irreducibly complex by nature, i.e. systems with properties we cannot say much more about other than that they are "emergent". Is the brain such a system? Or is the brain and other living systems just "complex pieces of technology" we do not understand yet.

To my mind there is no final resolution to this dilemma. Nevertheless, given our present state of knowledge, I would claim that there are many systems we should treat as if they are irreducibly complex, or where it would be irresponsible not to do so. Most social systems, I would say, fall in this category. Our understanding of any complex system is always from a certain perspective, and no definition can fix any such perspective as the final and objective one.

3. What is your favourite aspect / concept of complexity?

One of the originary questions of Philosophy is: "what is it to be human?" I have a keen interest in the way human beings give meaning to their lives, especially through cultural activities. Human beings transform the world they live in in all sorts of ways, both positive and negative, many of which are absolutely astounding. Some of the great works of art and science are simply beyond our comprehension.

One of the most beautiful aspects of complexity is that it provides us with a description of how such things are possible without referring to something supernatural. For me it has always been a little demeaning to ascribe a work of great revelation to some form of divine inspiration. At the same time, such a work also transcends the intentions and capacities of a single artist or thinker. Complexity theory shows us how we can encounter the most wonderful things, things beyond our wildest dreams, without having to invoke metaphysical commitments. Astonishment, laughter, surprise and love are all emerging properties of a complex world.

Given these considerations, it should not be too surprising that I consider the notion of "limits" to be one of the most important aspects of complexity theory. What are the limits of our understanding of complex systems, and what can we say about them? What is the status of our models of complex systems, and what is the standing of the knowledge we generate in this way? As humans we cannot know a complex system in its complexity. Knowledge is a human capacity, and we have to reduce the complexity of a system in order to be able to say anything about it at all. Knowledge of complexity is therefore always somehow limited. This does not imply that such knowledge is arbitrary, but it does imply that a certain modesty is required when we make claims about a complex world.

Our understanding of great artworks can also be partly understood in these terms. When we are confronted by a sublime work of art, our experience is exactly of having encountered our limits. We acknowledge that there is something which cannot be reduced to any mundane description. Acknowledging our limits is thus not a sign of weakness or ineptitude, it is an integral part of being human.

These aspects need further elaboration, also because it introduces a normative dimension into everything we do. If the presence of boundaries and limits is inevitable under complex conditions, then our descriptions cannot claim to be pure and objective. We make choices and compromises. The acknowledgement of these is what one could call the "ethics of complexity".

4. In your opinion, what is the most problematic aspect / concept of complexity?

The most interesting aspect of complexity would, of course, also be the most problematic. Thus, the notions of limits and boundaries, and the distinction between the two, remain very difficult to talk about. These concerns are also central to the work of Niklas Luhmann, but I have a certain unease with the strong influence of ideas based on the concept of autopoesis. A strong emphasis on the "operational closure" of complex systems is at odds with the central insight that complex systems are open systems which constantly interact with their environments in rich ways. The idea of closure is a necessary one, and I do not want to imply that Luhmann gets it wrong, but as a result of the ethical implications of complexity, I would hope that it is possible to generate understandings of complexity which allows for a more radical transformation of such systems through external influences, without denying the integrity which results from some kind of closure.

Ethical concerns are also central to what I see as probably the most problematic aspect of complexity theory at present: the continued positivistic claims made by many—most certainly not all—theorists who engage with complexity from a mathematical or computational perspective. I think particularly of the current importance being given to the so-called "power laws". It seems to me that these strategies remain reductive in the extreme. I recently witnessed a prominent complexity theorist, at an important event, claiming that everything which happens in society will be reflected in the Dow-Jones Index. This is not only an insult to so much of the difficult work done in the social sciences and humanities to understand the complexity of human interaction, it is also makes a travesty of so many insights from complexity theory itself.

This is not an argument against the use of mathematical and computational models in the least. It is an argument about the limits of any model of complexity. Since we live in a world already dominated by instrumental thinking and the values of the market, complexity theorists should be careful not to play into the hands of those who seek power through the control of the market. I would rather hope that a better understanding of complexity would help us to deal with some of the political, ecological and social disasters currently facing us. We can only do that by making the inherent complexity explicit, not by reducing vastly diverse phenomena to single indices.

5. How do you see the future of complexity? (including obstacles, dangers, promises, and relations with other areas)

I have already talked about the challenges and dangers involved with complexity theory, but a few things remain to be said about future developments. The first is to sound a cautionary note. I do not think that complexity theory, as computational modelling technique, will produce staggering new results. Similar to what happened in AI research, it will produce very useful things in a piecemeal way. It presents a different perspective and uses different methods which are vitally important, but it will not solve the big problems, just as we have not produced intelligent computers after more than 40 years of intensive research. This is not because of incompetence or fundamental errors, it is in the nature of the beast. The formal descriptions of complex systems can only address certain aspects of the complex human world, helpful as they are. Researches working in the discipline should be very careful about the claims they make about future possibilities.

My own understanding is, of course, subject to the same kind of qualification. It is possible that some real surprises are in store for all of us, and I will applaud them enthusiastically. Nevertheless, as I see things now, the greatest contribution from complexity lies not in its technological promise, but in the way in which it is influencing our understanding of the world. We should promote what can be called "complexity thinking", a style of thinking which is critical of claims based on reductionist thinking, yet at the same time mindful of its own limits.

One of the great benefits of complexity thinking is that it opens up a new interaction between the social and the natural sciences. This interaction should not imply the collapse of the one into the other, the difference between them should be maintained. Nevertheless, it provides a theoretical language in which the two can communicate in an interesting way. It should also lead to a greater mutual respect for the differences involved. There is no single understanding of complexity which ultimately trumps all the others. To a large extent the language of natural science is still seen as the final arbiter of truth. In order to change this it is of great importance that the human sciences develop a discourse on complexity which has integrity. In many ways our understanding of what it means to be human depends on it.

7

Jim Crutchfield

Director

Complexity Sciences Center, University of California, Davis, USA

---

1. Why did you begin working with complex systems?

Why I began working with complex systems is much clearer to me now than during the 1970s when I started to work on nonlinear dynamical systems. The way I think about my original motivations is best expressed as a historical reconstruction, giving some insight to the times and how the field developed. Hopefully, it casts the answer in a way that highlights the intellectual challenges.

From this perspective, one can see the recent history of complex systems as two sides of the same coin. The first side was the discovery that simple systems can appear random (and high dimensional), even when consisting only of low-dimensional, but nonlinear coupled components (aka "deterministic chaos"). Despite our intuitively negative reaction to unpredictability, deterministic chaos was great news at the time. Could much of the randomness that we see around us in the natural world be, underneath, hidden away, simple nonlinear dynamical systems? If so, we might be able to extract the hidden simplicity— states and dynamic—from a few, or even one, variable. Answering this question led to the introduction of the idea of reconstructing "Geometry from a Times Series" by Norman Packard, myself, Doyne Farmer, and Rob Shaw in 1979. By the mid-1980s that idea had spawned much work on nonlinear time series analysis, which provided an important push against the then-dominant focus on linear, stationary, independent, identically distributed processes.

However, success in analyzing how chaos arises set up a second, more problematic side of the coin—the attempt to answer a complementary puzzle. If simple systems can spontaneously generate random-appearing behavior, why do large-scale systems with many components appear ordered? In short, what is the origin of organization in a chaotic world? This question led me to explore the mechanisms of pattern formation in spatially extended dynamical systems by introducing the prototype class of map lattices and experimenting with video feedback, chemical oscillators, and ferrofluids. Quite a bit of groundwork on spatial systems had been laid decades before by statistical physicists studying phase transitions and critical phenomena. But their approach was not very "dynamical" and relied on assumptions of ergodicity to explore spatial organization through random samples of spatial configurations. To my mind, this use of ergodicity throws the baby out with the bath water, disregarding the question of the dynamical origins of organization. Exactly what mechanisms—state space structures—constrain and guide organization? Adding to the shortcomings with the statistical physics approach, spatial "structure" was often quantified only through correlation functions, which throw away almost all of what is unique and interesting in organized systems.

In looking back at this history, it becomes clear that randomness from simplicity and order from complication are two sides of the same coin and that coin was the concept of pattern and pattern emergence. So, in the historical evolution, the first intellectual innovation was to appreciate the mechanisms that led to randomness and the second was to understand the mechanisms that led to order. Today, we appreciate that much of what we call pattern, what we try to encode in theories and concepts, arises from the dynamical interplay of randomness and order. It's this middle ground that has been so hard to model, to predict, to explain. It fascinates us nonetheless, since it is in this middle ground where pattern (structural complexity) emerges.

So if you had asked me why I worked on complex systems in 1976, I would have only been able to answer that I was fascinated with chaotic dynamical systems: local determinism leading to long-term unpredictability; beautifully intricate attractor topologies; the rather deep concept of state (or configuration) space and a dynamic over it; and the like. I eventually came to call this "microdynamics" in the sense that one focused intently on very fine scale structures—homoclinic tangles, absolutely continuous invariant measures, uncountably many periodic orbits, fractals, self-similar basin separatrices, the spectrum of Lyapunov characteristic exponents... The origins of randomness were remarkably structured.

Exploring order and randomness has a long history and the first demonstration of deterministic chaos came through studying classical mechanics. The history goes back to the French mathematician Henri

Poincaré in the 1890s and his tour-de-force analysis of the three-body problem in mechanics which led, after a famous false start, to his articulation of the mechanisms of deterministic chaos. Since Poincaré, the study of nonlinear dynamics effectively departed from physics proper, being championed by the likes of Aleksandr Lyapunov, Balthasar van der Pol, Mary Cartwright and John Littlewood, Andrei Kolmogorov, Stanislaw Ulam, Edward Lorenz, Steve Smale, and many others in mathematics, engineering, and even biology. So, just as statistical physicists had studied "collective phenomena" for decades before the appearance of complex systems, the discovery of deterministic chaos itself had an important history—a history that predates much of statistical physics, in fact.

I'd like to think that what was different in my approach was realizing that something much more profound underpinned our study of each new nonlinear system than just the specific behaviors of each system. Despite rather active discouragement from senior colleagues, I became focused on generalizing studies of nonlinear systems to understand the very basic principles of modeling, that is, how we build theories, how we discover new patterns in nature.

So I can now say that I started to work in complex systems due to the realization that the focus of study should be patterns and how we discover them and not particular nonlinear systems in this or that application. To me, this is what complex systems is all about. And it is a very necessary activity, even if it smacks of a certain level of abstraction and philosophy. Acknowledging the latter, I came to call it experimental epistemology. The essential motivation for the research program was a simple, direct consequence of the most basic lesson of nonlinear dynamical systems: each nonlinear system requires its own explanatory basis. At root, we cannot blindly apply Fourier analysis, perturbation theory, wavelets, or any one of a host of accepted "complete" representations. We must understand each nonlinear system on its own terms. This realization brought up the question of intrinsic representation, how to learn new representations, and how to measure the amount of patterned-ness or structural complexity. With this professional epiphany I quickly moved on from nonlinear dynamical systems—showing that this or that physical or biological or social system could be chaotic—to the question of pattern discovery.

I just framed the question of complexity in a slightly more abstract way than most would—as the process of pattern discovery rather than analyzing specific examples of complicated systems. Nonetheless, this framing leads to two very concrete and clear questions. What is pattern and can we quantify the amount of pattern in a system? I believe answering these questions is what studies of complexity are uniquely about. Analyzing particular complicated systems and developing applications of complex system tools are crucial activities that bridge the general study of complex systems to the sciences and engineering. However, this kind of modeling activity survives healthily even in traditional disciplines and they end up recasting and owning the ideas, ignoring their origins in complex systems theory.

2. How would you define complexity?

Although I have framed "complexity" in a specific way, the term "complexity" is used in many, even contradictory ways. In particular, without an adjective qualifying the word "complexity", the question here induces confusion. Much of the controversy of what complexity is, if the issue really deserves such an exalted descriptor, has to do with folks looking to use a rich and important word in one and only one way—their way, without defining what they mean. Assuming they can define it for the problem domain that interests them, then all they have to do is use an informative adjective. A simple prescription, one would have thought. If it had been observed over the last twenty years, much redundancy and reinvention could have been avoided and we would probably be further along.

Putting superficial semantic confusions aside, I will answer the question by rephrasing it.

First, there are senses of the word that are now seen to be in tension. Many dictionaries give two definitions for "complexity". Definition 1 is "complicated", "noisy", "random", and so on; Definition 2 is "sophisticated", "intricate", "consisting of related parts", "structured", and the like. The first means "without organization" and the second, "replete with organization". And, worse, there's even the limit in which the latter turns into the former: as one dictionary puts it "complicated in structure". So we can see why some confusion was inevitable: our very reference documents conflate distinct properties.

Second, the mathematics of randomness is fairly well understood, thanks to such luminaries as Andrei Kolmogorov, Richard von Mises, Claude Shannon, Alan Turing, Ray Solomonoff, and Gregory Chaitin. Thus, I do not use "complexity" in the sense of Definition 1; rather I use "random" or "unpredictable" to describe this property of a system. Instead, I use "complexity" to describe the properties of a system captured by Definition 2. These choices are a simple, personal shorthand. It seems like a waste and redundancy to take a rich word like "complexity" to mean randomness. "Random" is a perfectly good label for its referent. Nonetheless, a confusion could arise with a shorthand and so when trying to be completely clear I say "structural complexity", trying to adhere to the adjective dictum above.

So, finally, the question becomes, What is structural complexity? This I can answer unambiguously and concisely: Structural complexity is the amount of historical information that a system stores. Without recounting the technical results, the upshot is that there is a deep connection between how a system stores and processes information—how it "intrinsically computes"—and how the system is structured.

Thus, within the computational mechanics framework I use, complicated systems are not necessarily complex. One has to show that a complicated system of interest actually stores and processes lots of information. Then it is structurally complex.

3. What is your favourite aspect / concept of complexity?

I like that one can use complex systems to understand, with a very small set of ideas, much of the natural and human-made world. In a nutshell these ideas come from studying the commonalities between dynamical systems theory, statistical mechanics, information theory, and computation theory.

They consist of (i) the geometric view that dynamical systems theory gives of the state space (and the structures there that drive and constrain behavior), (ii) the notion of structure captured in the theory of computation, and (iii) the concept of information (aka entropy in statistical physics) from communication theory. Pretty much everything else in complex systems is one or another solution strategy, algorithm extension, or optimization trick that builds on these three foundations.

The interest and fascination in discovering patterns in our world is as old as humanity. Now that we have new mathematical and computing tools appropriate to complex systems, we are in position to make profound progress in understanding natural and, even, engineered systems by analyzing their behaviors and organizations.

We should not, though, get too far from the intellectual innovations that led us to this point. There are many important lessons and even a sense of inevitability when one looks at the historical momentum behind the questions that complex systems pose. In particular, the recent interest in complex systems is, in many ways, a revival of Wiener's push for cybernetics in the 1940s and 1950s. It's not too much of a simplification, and perhaps it does some honor to him, to see complex systems as "nouveau cybernetics", as a rekindling of his and others' enthusiasm during that period. Unfortunately, the word "cybernetics" has a checkered past. For example, it is very hard to see cybernetics in contemporary computer science, though that field's founders played key roles in articulating the problems of complex systems.

An important, perhaps under-appreciated, aspect of complex systems is its novel contributions to scientific methodology. One of these you might call emergence analysis which attempts to address the question, By what mechanisms did behavior X or structure Y appear? This is fundamentally different from how questions were posed previously. Let's take an example from population genetics. It has been observed that genomic mutation rates vary over time and from gene to gene. One way to model this is to add a "modifier" gene that directly controls mutation rates. This gene too is subject to mutation and so one can analyze how mutation rates vary by solving the population dynamics problem of whether the modifier gene persists or atrophies as a population of genotypes evolves. This approach is very different from understanding why and how a modifier gene might appear in the first place. What improvement in survivability would lead evolution invent a gene that modulated the rate of change of other genes? How could such an innovation be encoded? This is a question of evolutionary innovation—the evolutionary analog of pattern discovery. Conventional population modeling assumes a specific sophisticated mechanism, whereas treating the question as a complex system allows one to investigate conditions promoting evolution of diverse sophisticated mechanisms.

Another of my favorite aspects of the research process for complex systems is that is it is very "geometric" and so, to me, visual, if you are willing to buy into a certain level of abstraction. That abstraction is the state space—a representation of the set of all possible configurations a system can be in. One's understanding of the emergent patterns in a complex system is expressed by delineating the geometric structures in the state space which lead to the system's behavior.

Over the years, my desire to view state space led to repeated searches for new kinds of exploratory tools. First, my graduate student colleagues and I at UC Santa Cruz used analog computers (this was the 1970s) to solve differential equations with chaotic solutions. We also made 16 mm films of, for example, the cross sections of strange attractors to understand their intricate topologies.

Rob Shaw and I made a reel-to-reel video (1970s!), now apparently lost, of the chaotic dynamics and bifurcations in a dripping faucet. I also developed an experimental system for interactively exploring spatial pattern formation. This was based on the reaction-diffusion dynamics supported by an electronic-optical computer, commonly called "video feedback". These instruments were special purpose and speak rather directly to the technologies of their time. Today, of course, technological circumstances have changed immensely; we now regularly use massive Linux clusters for multiagent simulations and even immersive visualization (see KeckCAVES.org) to explore complex systems.

Let me finish with a simple, personal favorite about complex systems. I enjoy the excuse that it gives for exploring different fields of science and the arts. This could very well be a product of my having an attention span of only about five years. But it is particularly rewarding to comparatively explore a general concept like emergence or pattern by delving into music, physics, psychophysics, biology, bioacoustics, and social science.

4. In your opinion, what is the most problematic aspect / concept of complexity?

Early studies of complex systems became caught up in a difficult social change in the sciences in the 1980s and 1990s: increased public attention that led to increased self-consciousness. (This is not really unique to complexity, but it did play a key role in its development.) Never had there been so much interest in books on science and magazines devoted to it. One or two key books in the mid-1980s seem to have woken up publishers to a public hunger for news on science and technology. One manifestation of this was the hugely increased role of science journalism in the late 1980s.

There were many positive benefits to this. One was that your parents and relatives were much more appreciative of what you did, even if they admitted to not really understanding it. More broadly, the public and, very importantly, young students received a much more lively picture of contemporary science and mathematics. Chaos, fractals, emergence, and the like were all fantastic fodder, conceptually and graphically, for an intelligent lay audience.

The downside, of course, appeared as changes within the science community, including the occasional lack of warmth with which one's immediate colleagues greeted one, when a result got widespread press.

In the 1980s and before, getting exposure through the press was largely new to most areas of science. In physics, anyway, people were downright skeptical of this shift. (We were faceless workers collectively building the edifice of science, through which society benefited and so paid our daily wages, right?) This was particularly true when journalists, desiring to interest a broad readership, would focus news articles on personality and lifestyle. As a journalistic strategy it was hugely successful and, for better or worse, now twenty years on, it is common practice; even one with which hard-nosed physical scientists seem to have come to terms.

The general area of complex systems was a prime beneficiary of this more-public exposure of science. Even if universities hesitated to accept changes to their curricula to teach complex systems, funding agencies, industry, and even private philanthropists were supportive. Without them, the field would be markedly much less advanced today.

But it also led to a level of self-consciousness that had a chilling effect on cooperation and the development of shared goals. A new kind of entrepreneurial attitude developed, somewhat spurred on by the increasing sophistication of computing technology. Computers, once only unacknowledged handmaidens to science, came to be seen as essential tools. In many cases, they were the only way to access and study many problems in complex systems. However, those tools had to be programmed, which was time consuming, very difficult, and expensive. So, if you could craft a tool that reduced the difficulty of modeling and simulating complex systems you could control the direction of science. More to the point, a useful tool was a product that could be sold by companies who determined the content.

Fortunately, this entrepreneurial period has stabilized and, many thanks to the emergence of global scientific collaboration via the Internet and the Open Source movement, the science of complex systems (and many other areas) is no longer hostage to the commercial entrepreneurial spirit. The upside, as one looks to the future, is an even greater rate of discovery and innovation as tools are collectively developed and as the tools' source code is there for all to see and to improve much more quickly.

So was this increased self-consciousness really a problem? It is hard to know. Perhaps the long view is that social dynamics, even in science, moves to the middle ground I described above, becoming truly complex as a result, as it swings from one extreme to another.

5. How do you see the future of complexity? (including obsta-

cles, dangers, promises, and relations with other areas)

I'm very concerned about training and education in complex systems. I'm happy to see many of us developing books and specialty courses. However, I'm concerned that, given the central and broad importance of complex systems, there are very few appropriate multiyear graduate training programs. I've been working in this area for over thirty years now and, in that setting, I feel we, or at least I, have failed our students. To my taste, there has been too little synthesis and too much competition. The result is a dilution of the original spirit and insight. (How ironic for a field one of whose central goals is to understand emergence.) One central cost has been a lack of cooperation to build training programs.

On the up side, much progress has been made in the research arena. Now is the time to take the (perhaps substantial) effort to rework those results so that students can be introduced to the concepts systematically. The goal is that they learn deeply enough to extend them creatively to attack the many remaining, truly complex problems.

So, it's time to move on from "complexity". The heydays were in the early 1990s. Much progress was made; though, much of that progress is still being digested.

The label "complexity" was positively useful then. It was vague enough that people in very different disciplines working on their field's hard problems could come together with a hope of learning something from others in other fields. They came with open minds and enough energy to get through the difficult disciplinary boundaries of vocabulary and research style. The label was also specific enough that an interest in "complexity" was an effective filter that kept more discipline-focused scientists at bay.

At this point, however, the word has entered a stage of overuse and so its meaning and utility have been diluted.

This is not to say that the problems that have fallen under this rubric are less interesting or less important. In fact, just the opposite. We have never faced problems of the level of complexity, either natural or of our own making, than we now do. For example, the age-old problem of individual action versus collective function is more present than ever. We see this in both the natural world, as affected by us, and in the artificial world, as engineering innovations allow us to assemble ever-larger and increasingly vulnerable socio-technological systems. I believe that the concepts and tools of complex systems will be key to understanding and solving such problems, including sustainability, social justice, and economic stability.

However, the field is a victim of its own success. It's now time to start making finer distinctions than simply saying that this or that system is "complex". How exactly is it complex? How much information processing does it do? What is the minimal dimension it lives in? What are the intrinsic coordinates for that space? What are the effective forces that drive its behavior? At what levels are different kinds of information processing embedded? How does collective function emerge from individual behavior? Can we design policies for individuals that lead to desired collective outcomes?

By definition, complex systems are some of the hardest problems to approach scientifically and mathematically. Frankly, I feel that modern mathematics, despite the amazing insights of our predecessors, is currently not up to the task, specifically the problem of representation. Very few nonlinear problems can be solved in closed form; that is, the underlying mechanisms and solutions cannot be expressed in mathematically tractable ways. We need a very new approach.

People today say computers can fill the mathematical void. I would agree with this only partially, since powerful computers can be programmed to simulate models that are as complicated as many natural systems. Beyond the conceptual hygiene that comes from porting the mathematics of an idea to a computer language, often little is gained. More to the point, having petabytes of simulation data is not the same thing as understanding emergent mechanisms and structures. We need to understand mechanisms and structures to build predictive theories—theories specific enough to be wrong. Perhaps even more troubling is the recurrent conflation of writing a 20,000-line LISP program as a model of machine intelligence and claiming that that program means you understand how intelligence works. This is far from the case, since such a program most likely is very complicated itself. In any case, it does not directly represent the emergent mechanisms that lead to its behavior. Without emergence analysis of the running program, for example, one can't say which portions interacted cooperatively to produce a successful solution to a learning task. A useful comparison is a program that simulates the famous Logistic Map, which need be only several lines of code long. It has taken decades to understand the rich chaotic behavior in that nonlinear system.

Very recently, this line of thinking reached a new nadir with marketing successes, in targeting online advertising, that rely on mining large empirical data sets. The successes led information technology leaders to claim that theory is no longer necessary. However, in devaluing theory the pure-simulation, pure-data, and pure-computing approaches sideline scientific understanding. Complex systems provide no better antidote to such thinking. We live in a very prosaic age as these examples show, nonetheless relying heavily on the conceptual insights of past great, innovative thinkers. To devalue theory is to preclude the future reduction-to-practice of today's conceptual insights. In light of this, I've often joked that my role is to make the world safe for theory. At the largest scale, it's really an ecology of pattern discovery. We should move away from the rhetorical extremes to the dynamic synthesis found in the middle ground.

These things said, I'm very optimistic about the future of the intellectual momentum that "complex systems" represents. There are many interesting and absolutely fundamental problems that, on the one hand, we don't understand and that, on the other, the new methods and thinking are ripe to solve. I hope to look back to this time and marvel at how naive we were.

8

Bruce Edmonds

Director

Centre for Policy Modelling,
Manchester Metropolitan University Business School, UK

---

1. Why did you begin working with complex systems?

My answer to this is rather personal. It starts with a very basic fact, namely that my father was a physicist and my mother a social worker. I remember them discussing some of the social theories my mother studied. My father thought little of these theories, saying that knowledge is no good unless you can state it precisely (i.e. not only in vague analogies) and it gives you some "leverage" upon the world (i.e. it is useful in some way). My mother's reply was that social matters were much more complicated than physicists imagine and not explainable merely in terms of atoms and forces. The argument was never settled—neither had a good reply to the other's points. The reason turns out to be that they were both right, but it took me many years to realise this.

Other results of this parentage was an interest in social issues, which my mother talked about, and being brought up with computers which my father brought home from his laboratory. At some stage I read a book about mathematics, and was fascinated. In particular it listed an axiomatisation of set theory, explaining that all known mathematics could be expressed in the set theory that they specified. This interest (along with a complete failure to succeed in writing good English or learn any facts) led to me studying my mathematics as my first degree. Then however, I went into youth work, partly because: I found social systems more interesting and I had become aware of some of the limitations of analytic mathematics (in particular its applicability).

By 1992, I had been thinking about several related things for a while, namely: the limits to formal analytic modelling techniques; the difficulty of understanding social systems; and what exactly it is that makes something complex. For my sins I started a doctorate in philosophy on "the meaning and definition of complexity". At that time there were relatively few things written about the concept so it gave me a good excuse to read material from almost any part of the library. In 1994, by sheer dumb luck, I got a job with Scott Moss who was using computers to model economic systems. I started doing this as well, as part of what later came to be known as "agent-based social simulation": using complex computer programs to try and understand complex social phenomena. Despite my hubris in choosing the topic and a ballooning list of references I managed to complete my thesis in 1999.

Thus, although I first came across complex systems in an entirely abstract way, I ended up rejecting general abstract approaches and instead have concentrated on practical ways by which we can seek to understand them using simulation modelling.

2. How would you define complexity?

The nearest I have come to this is the definition which is the conclusion of my thesis, namely:

> *Complexity is that property of a model which makes it difficult to formulate its overall behaviour in a given language/framework, even when given reasonably complete information about its atomic components and their inter-relations.*

The essential aspects of this are that:

- you will only get a more specific definition of complexity given specific contexts, fields or frameworks, there is no general approach that is practically applicable;

- complexity usefully appertains to models of phenomena rather than to the phenomena themselves, it is a property of the models and only the phenomena if you conflate your model with the reality it represents;

- as projected upon the world complexity is a negative concept: covering everything that is not simple, thus almost anything can be thought of as complex.

For details see my thesis [1]

3. What is your favourite aspect / concept of complexity?

The complexity bandwagon has helped encouraged some *existing* trends in the development of science, including the following.

- The use of complex simulation models instead of (or along with) analytic models, thus adding to the menu of tools available to the scientist. No longer is it always felt necessary to "shoe-horn" phenomena into analytically tractable mathematical or statistical models when this necessitates the use of assumptions that obscure important aspects of what is being understood. In particular one does not have to use numerically based models but can model much phenomena in a more straight-forward manner. This has resulted in a swath of simulation models that are more specifically descriptive in nature and do not resemble a traditional theory from physics in that the model itself can be difficult to understand completely.

- A re-thinking of the purpose and processes of science. In particular the range of uses that models can be put to (for example to inform and be informed by good observation), as well as different ways of using models together (e.g. in chains of models or as complementary to each other). Simplistic accounts of "how one does science" have become less narrow and prescriptive.

- The use of a wider range of evidence. For example, in computational models of social phenomena it is possible to utilise reports from people of what they do and why by including this process in a formal computational process. Thus the introduction of simulation models allows for more of the evidence to be formalised and thus seriously considered as part of the scientific discourse. Science now tries to deal with a broader range of evidence (and by implication phenomena) than it did previously.

- A long-overdue breakdown of the myth that the truth about our universe must be, in some sense, simple. Thus the excuse of "for the sake of simplicity" is gradually being replaced by more honest phrases referring to limitations of time; computational

---

[1] Joke!

resources; or imagination. No longer does everybody expect the truth to be simple, nor are they only convinced by accounts that are simple. In this respect science is growing up, with some acceptance that many fields (e.g. economics) will end up looking more like biology and less like physics.

4. In your opinion, what is the most problematic aspect / concept of complexity?

Since complex systems cover all systems that are not simple, it includes pretty much everything we encounter. Under this usage, "complexity" rapidly looses any useful positive meaning and becomes a "dustbin" concept, rather like "context" or "system". For this reason there is not, and will never be in any meaningful sense, any "science of complexity"—a science of complexity makes no more sense than a science of non-red things. Similarly (at least so far) there is no coherent body of knowledge that could be honestly called "complexity theory" but rather a collection of techniques and tools from different fields, loosely (and sometimes rather uncomfortably) bunched together under the same label. The hype associated with these terms confuses the public and raises false expectations within funding bodies.

Thus, in my view:

- There is no hidden principle of complexity to be found behind observed phenomena;

- There never will be a "Science of Complexity";

- There is no "Complexity Theory".

5. How do you see the future of complexity? (including obstacles, dangers, promises, and relations with other areas)

It has no future as an identifiable field or cluster of fields.

Like Systems Theory or Cybernetics before it, it will slowly fade away and across into the humanities and public discourse. Simulation approaches will take its place alongside statistical and analytical approaches as "just another tool" to be used as and when it is helpful. However, some of the lessons the label stands for (e.g. those listed in answer 3 above) will permeate all areas of science and become part of the accepted or standard view.

Thus the dangers are short-term and common to many other new trends and labels. Approaches associated with complexity will be subject to too much hype for a while and their usefulness will be both under- and over-estimated, depending on the age of those who judge them. While this stage lasts, there will continue to be much confusion caused by the word "complexity", so much so that serious researchers will start to seek to avoid using it. On the other hand politicians will start to use it in speeches, demanding such as "a complexity-led solution" to particular problems.

It is a flash-in-the-pan, but it signals slower and more fundamental changes in the way science works, as science continues to adapt to the subject matters it can cope with.

9

Nigel Gilbert

Professor of Sociology
University of Surrey, UK

1. Why did you begin working with complex systems?

During the 1980s, I had been involved in a number of projects—
ranging from designing speech understanding systems to developing
interactive computer programs to help people claim welfare benefits—
all of which involved trying to import some of the knowledge and ideas
of social science into the design of information technology. After some
ten years of this, I thought it was time for some pay-back to social science and wondered how the technology could help with the advance
of social science. Around the same date, Jim Doran, a professor of
Computer Science at the University of Essex and a keen and knowl-
edgeable archaeologist, invited me to collaborate with him in building
a model of the 'Emergence of Organised Society' in Palaeolithic Europe. Archaeologists had two somewhat different theories about the
transition during the last ice age from a hunter-gatherer society organised around small family groups to much larger and more complex
societies (one trace of which are the famous cave paintings of south-
west France). I was intrigued by this idea and readily agreed to help.
Jim and I spent a couple of years developing a Prolog program running
on a Sun workstation that simulated the emergence of leader-follower
relationships in environments of resource scarcity.

We would nowadays call this an agent-based model, although to
us then, it was just a object-oriented logic program. In retrospect,
our plans were far too ambitious (I am not sure that even today it
would be easy to fulfil our design goals), but the work did suggest that
developing 'social simulations' might be both interesting and offer op-
portunities to advance social science. Thinking that it would be useful
to see what others made of this idea, Jim and I organised a workshop
in Guildford, UK, in April 1992, under the title, Simulating Societies.

We said in the Call for Papers that "Although the value of simulating complex phenomena in order to come to a better understanding of their nature is well recognised, it is still rare for simulation to be used to understand social processes. This symposium is intended to review current ideas on simulating social processes, compare alternative approaches and suggest directions for future work." When we drafted this text, I knew that we were running a risk: I knew of very few people doing anything like this and it seemed quite possible that no one would come. However, some 24 participants showed up from all over the western hemisphere and the meeting was a very successful, creating a community of interest which grew rapidly from that small beginning.

The papers from that meeting were published in Simulating Societies and were followed by a collection from a subsequent workshop in 1995, Artificial Societies. Other similar, small scale meetings were held, mainly in Europe, and gradually began to define a common set of assumptions and methods. At this time we were still talking about 'distributed artificial intelligence', but gradually we changed to 'agent-based modelling' (or agent-based social simulation, or multi-agent based simulation—attempts to define and distinguish these terms used up much energy). Most of the meetings generated sets of conference papers that we wanted to get published, but most conventional disciplinary journals would not touch this mysterious stuff that used concepts and techniques remote from the mainstream. I eventually ran out of publishers willing to take on yet another edited collection on social simulation. The solution, I thought, was to start a journal, but publishers were not much more enthusiastic about that than about the edited collections I was offering them. They assumed that the material would appeal to a very wide interdisciplinary audience, but this, far from being an advantage, was a marketing nightmare: to whom should they send their brochures advertising the new journal? Without a clearly defined target market, they could not work out a marketing plan. I thought that this was the end of the road for social simulation, since publication seemed to be getting impossible, until I realised that a recent project to create an online journal for UK sociology (Sociological Research Online) could be adapted for our purposes.

To considerable scepticism from many of my social simulation colleagues, we started the Journal of Artificial Societies and Social Simulation in 1998. This is only available online—there is no hardcopy version—and it is entirely free, to both readers and authors (no subscription and no page charges). At first, we decided that it must be free because we thought that we would never be able to get readers if we charged. This was possible because we could host the journal without charge on my University web server, we had no printing or mailing costs and all the editorial work would be done by the editors themselves. As always, the first few issues were the most perilous, and I often worried about how I was going to fill the next issue, but we managed. The biggest obstacle was the reluctance of researchers, especially in the US, to contribute because they thought that an online journal publication would not count towards their tenure. Nowadays, JASSS is included in the Science Citation Index and has a high impact factor for a social science journal, so this is less of a problem, but it is still the case that the more conservative academic establishments regard JASSS with some suspicion. On the other hand, publication in JASSS reaches a much wider audience than could be achieved with hardcopy publication: around 100,000 articles are viewed per month (compared with the typical successful journal print run of no more than 1000 copies).

2. How would you define complexity?

Complexity crept up on me unawares. Everyone knows that societies consist of many individuals, interacting and changing, and that people are different, one from another. But the idea that a society could be regarded as a 'complex system' and that concepts from other disciplines might illuminate how societies emerge, develop and fail was one that took a long time to reach me and, it would be fair to say, still leaves the great majority of sociologists cold. One obstacle is the very notion of 'system': the concept has an infamous history in sociology, associated with figures such as Talcott Parsons, whose idea of a social system was one in which there is equilibrium and an internal unity of purpose that was later rejected as ideologically biassed and empirically false. While systems theory achieved a new lease of life in Germany under the influence of Luhmann who was much influenced by complexity theorists such as Maturana, most sociologists in the English speaking world have rejected anything that looks like systems theory in favour of approaches that start with individual action and the social construction of meaning.

Against this background, the application of ideas of complexity imported from physics tends to arouse suspicion among my colleagues. On the other hand, an approach to the analysis of social processes based on complexity does help to unravel some classic conundrums, such as whether individual action or social structures have precedence and how they are related (the so-called 'micro-macro link'). Complexity provides a vocabulary of emergence, attractors and dynamics that helps to dissolve these conceptual problems by showing how behaviour can lead to structure and structure can constrain behaviour. One of the main themes of my own work has been to show how the ideas of complexity need to be adapted to analyse human societies: for example, that people can themselves detect and react to emergent features of their own society, in a way that inanimate matter cannot: this leads to the idea of 'second-order emergence' or 'immergence'

3. What is your favourite aspect / concept of complexity?

The critical aspect of the complexity approach for me is that it deals with dynamical systems. Human societies are constantly developing and changing and a theory that fails to recognise this or (as with some perspectives in economics) assumes that societies are in equilibrium, is completely unsatisfactory. Many sociological frameworks start from the position that society is constructed or (in complexity language) emerges through interaction and this is also the position of complexity science.

4. In your opinion, what is the most problematic aspect / concept of complexity?

The other side of the coin is that complexity science, by aiming to be a general theory, tends to assume that people are really not much different from atoms or bacteria. It is true that in some limited circumstances, people do react in the approximately the way that particles might. For example, when one is rushing down a crowded street with the goal of reaching one's destination as quickly as possible, and with no distractions and no one accompanying you, your motion can, it seems, be approximated quite well by a particle controlled by a few simple rules. Similarly, models based on a few reactive rules seem to work quite well in predicting the flow of cars driven down highways. It is somewhat more questionable whether simple models can be useful in understanding changes in political opinions, for instance.

5. How do you see the future of complexity? (including obstacles, dangers, promises, and relations with other areas)

Within the social sciences, the complexity approach is certainly spreading and becoming more influential. It is already significant within economics and important in geography. the major obstacle to a further advance is the difficulty of finding literature that is accurate, relevant and comprehensible to social scientists who do not have an advanced mathematical training. However, suitable texts are gradually emerging. The danger is that the mathematics will be short-circuited, that complexity will be used merely as a loose metaphor and that its concepts and ideas will be lost in a miasma of 'hype' and the approach will lose any respectability. One way of avoiding this is to link with physicists, biologists and ecologists, in inter-disciplinary collaborations, in which the various concerns and perspectives on complexity can be integrated, but this involves a great deal of goodwill and understanding of the others' points of view that not all social scientists are prepared to undertake.

10

Hermann Haken

Professor Emeritus
Institut für Theoretische Physik, Universität Stuttgart, Germany

---

1. Why did you begin working with complex systems?

For me it is very hard to answer this question in a proper way. The reason lies in the fact that all my scientific life, I have been working on what is now called complex systems, though I haven't been aware of this fact for quite a long time.

The fields I have got involved are as follows:

A. Mathematics

My Ph.D. thesis in mathematics was concerned with the "word problem" of group theory. The problem is rather easily described. Think of a string of symbols, such as letters A, B, C, etc. which can be arranged in all sorts of sequences, where the same letter may be used several times. Now from such sequences select some specific sequences and put them equal to one. These are the "defining relations". Then the question is: using simple rules of group theory concerning the multiplication of symbols, are two expressions equal when the defining relations are applied. Consider the example of a defining relation: $AB = 1$.(*) Are A^2B and A equal? Yes, because of (*). $A^2B = A(AB) = A$. Are A^m and B^n equal, $m > 0, n > 0$? No, because due to (*) $B = A^{-1}$. This example is deceptively simple. But when more complicated defining relations are admitted, the problem does not only become formidable, it cannot be solved in general. In other words, there is no general algorithm (procedure) by which one can decide in a number of *finite* steps, whether two "words" are equal. As it turns out, there is a deep relation to the halting problem of the Turing machine. Without knowing it at the beginning I had run into a really complex problem (actually I could solve this problem for certain classes of defining relations).

B. Solid State Physics

Later, I switched to solid state physics. Consider a crystal, e.g. a salt crystal (NaCl). Macroscopically seen, it looks simple. When we analyze its structure, it has a specific crystal structure, which is periodic and thus again looks simple. When we start studying the dynamics of its electrons, the problem becomes complex. In semi-conductors and metals we are dealing with the so called many-body problem. The interaction between the numerous electrons has to be properly taken into account. These interactions may give rise to surprising effects, e.g. to super-conductivity. At sufficiently low temperature in some ring shaped metal, an electric current once produced never stops. Actually, while in some classes of super-conductors, the mechanism of super-conductivity is well understood by means of the Bardeen-Cooper-Schrieffer theory, more recently discovered (high temperature) super-conductors are still waiting for their explanation. I myself got involved in the problem of super-conductivity basing by work on the Froehlich model of super-conductivity.

Another effect in solids occurs in semi-conductors. Conventionally, we assume that the electric current in semi-conductors, such as silicon, is carried by the negatively charged electrons. However, there are experiments that seem to indicate that electric current can also be carried by positively charged particles, the so called "holes". Heisenberg had shown how the existence of holes can be traced back to that of electrons, using sophisticated methods of quantum field theory. I went a step further, namely I studied what happens when an electron and a hole come together and interact with lattice vibrations. It turned out to be a complex problem because of the numerous degrees of freedom and the nonlinear coupling of lattice vibrations to the particles. The interaction between the hole and electron with the lattice vibrations changes the direct interaction between hole and electron. The direct Coulomb interaction between hole and electron is weakened. On the other hand, my theory showed that when the particles carry the same charge, an attractive interaction results. This is actually the mechanism on which the Bardeen-Cooper-Schrieffer theory explains super-conductivity. I should mention that the formation of electron pairs, the so called Cooper pairs, was derived differently by Cooper, who took into account the effect of the so called "Fermi surface".

C. Lasers and Quantum Optics

The third field, where in the beginning I fully unconsciously started working with complex systems was the laser, a by now well-known light source, that may produce very intense coherent light. I was led to this problem of the laser by some lucky circumstances. In the spring of 1960, I spent several months as visiting scientist at the Bell Telephone Laboratories. The younger generation may not be so well aware of these labs that played a leading role in solid state physics in the decades, starting with post-war science. I soon learned from my friend Wolfgang Kaiser, who had been working for several years at Bell Telephone Laboratories that they were working on a totally new kind of light source, which was called the "optical maser". Before I arrived at Bell, I even haven't heard what a maser is. It is actually a device that allows one to produce coherent electromagnetic waves with a wave length of millimeters to centimeters. The maser principle had been suggested by several groups, but I don't think I should enter this historical debate here. In 1958, Schawlow and Townes suggested that it may be possible to extend the maser principle into the optical region, i.e. to produce *light waves* based on the maser principle. "Maser" actually is an acronym for "microwave amplification by stimulated emission of radiation" and the research was concentrated on realizing such an amplification principle. Actually, as I was the first to show later, the laser does not act like an amplifier, but rather as an oscillator. But this is the story I am going to tell now.

At the time when I was at Bell labs, two groups started there to realize the laser (as it was called later as "light amplification by stimulated emission of radiation"), namely the group by Kaiser and Garret on solid state devices and another one by Ali Javan and others on gas lasers. So I had intense discussions, both with Wolfgang Kaiser and also with Harry Frisch and we conceived quite a number of cavity arrangements, i.e. spherical active material into which a cone was drilled, a device that actually many years later had been realized.

In how far is the laser a complex system? At first sight, the device for the production of laser light is quite simple. For example, in the case of a gas laser, it consists of a glass tube filled with atoms or molecules. At the end faces of the glass tube two mirrors are mounted, one of them being semi-transparent. The individual molecules are excited by an electric DC current, sent through the glass tube. By means of the collision with an electron, a molecule may be excited and then may emit a light wave. So, after the collisions, individual light tracks are emitted. Since the emission acts were quite independent, the light thus produced may be visualized as being composed of Spaghetti. If light were audible, it would sound like noise of the sea. The idea of the mirrors was lent from maser technology. Here the microwaves were produced in a metal box that allowed for the existence of only waves with discrete frequencies (or hopefully only one). In the case of the laser requirements were more modest. In the first place the mirrors served the purpose that those waves that were running in axial directions, were reflected sufficiently often, so that they could interact with the molecules intensely. Now a crucial effect appears for the realization of light amplification, namely the process of stimulated emission, originally introduced and postulated by Einstein. It is better to think in terms of the so called "photon picture" in which light is visualized as being composed of individual particles, the photons. Thus, when several photons are already present, by means of stimulated emission, their number is enhanced. Now we arrive at a picture that was developed in the early stages of laser theory. Namely with increasing pump strength, i.e. level of excitation of the molecules, more and more light waves are emitted. Those waves whose frequency is closer to resonance of the molecules, are amplified more strongly and have a longer lifetime in the glass tube, so that the whole emission line profile becomes sharper and sharper. This was the desired effect of *Line Narrowing*. This was the generally accepted picture, before I started my own work on laser theory. Here I proceeded in two steps. In 1962, I developed what later should be called a "semi-classical theory" of the laser. In it, the internal states of the molecules or impurity atoms in a solid state laser were treated quantum mechanically while the laser field was treated like a classical electromagnetic field. As it turned out, at the same time Willis Lamb developed a similar theory, which he then published in 1964, while our publication was done in 1963. This theory showed that there is a sharp threshold for the occurrence of laser light. Below the threshold there is no light emission at all, whereas beyond that threshold that is determined by the energy pumped into the laser, a coherent wave appears. This theory predicted the existence and competition between different kinds of laser waves, and frequency displacements because of the nonlinear interaction between laser oscillations. I elaborated my theory jointly with my diploma student H. Sauermann.

But to me, there remained a deep puzzle, namely according to this theory, below a threshold there should be no light emission at all, whereas beyond it the coherent laser light emerges. This is, of course, in contradiction to all experimental facts, because light emission starts from the very beginning, once the molecules are excited. Thus the semi-classical theory of Haken-Sauermann /Lamb could reproduce only half the truth. To resolve this puzzle, I had to introduce the fully quantized light field. So in a first step I derived fully quantum-mechanical equations which contained the quantum dynamics of the electronic states of the molecules, the quantized light field and the interaction between these two systems. However, this was not enough, because, as we had known from the semi-classical equations, the electronic degrees of freedom, as well as that of the laser field, had still to be coupled to so called heat baths or reservoirs, which give rise to damping, and, what was especially important, to quantum-mechanical fluctuations. Based on a representative equation of such a type, I could show in 1964 that laser light undergoes a dramatic transition from the incoherent state below threshold to the fully coherent state above threshold. I calculated all the typical features of such a transition, such as critical slowing down, critical fluctuations, the coherent state, symmetry breaking, etc. In other words, I had shown that this laser transition shows just all the typical features of a *phase transition* known from systems in thermal equilibrium. Thus, laser light was the first example of a phase transition, actually of a second order phase transition, of a system *far from thermal equilibrium*. In the context of complexity theory the following feature seems to be particularly interesting: in spite of the fact that more and more energy is pumped into the laser system, so that one would expect more and more irregular light emission, eventually the system settles down in a *highly ordered* state. Thus a laser is a wonderful example of self-organization. I think that is an account of my own involvement in dealing with complex systems and especially the exploration of spontaneous formation of ordered structures in systems away of thermal equilibrium. What I think is nice about this approach is the fact that, starting from first principles from the quantum-mechanical level, one can derive the structure formation of laser light in every detail. The theory has been refined later in a variety of ways, where e.g. it could be shown by Graham and myself that in an infinitely extended laser, a laser light distribution function can be derived that corresponds to the Ginzburg-Landau theory of super conductivity in a one to one fashion. These profound analogies indicated to me that there must be a whole field of research in which the spontaneous formation of structures in systems away from equilibrium is studied, especially close to phase transition-like points.

D. Synergetics: Science of Cooperation

Thus in a lecture given 1969 at the University of Stuttgart, I coined the word "Synergetics" in order to characterize a new field of research that deals with exactly these phenomena but not restricted to physics but in all sorts of fields. These fields include also biology. Just to mention an example: There exists a profound analogy between laser light formation on the one hand and the development of species of prebiotic molecules as described in the theory by Eigen and Schuster. Also W. Weidlich had found phase transitions in models of the formation of public opinion. Thus a general search for non-equilibrium phase transitions in a great variety of systems and for spontaneous structure formations seemed to be well justified. Actually, the systems belong to quite different fields, ranging from physics, chemistry, biology, over sociology, economics to medicine, brain research and psychology.

E. Brain Dynamics

The human brain is certainly the most complex system we know of. Nevertheless, I found it worthwhile to apply the concepts and mathematical tools developed in Synergetics to a number of concrete problems of brain research, such as movement coordination, pattern recognition, decision making.

I think this rather extended statement of mine should be sufficient here for answering the question why I began working with complex systems.

In the beginning, i.e. some 50 or 40 years ago, I entered this field of complex systems (how it is called now) quite unconsciously. But when I initiated Synergetics, my approach implied the systematic study of complex systems. This is witnessed, e.g. by the preamble of volumes of the Springer Series of Synergetics.

To me the basic question is: How far does "complexity theory" go beyond Synergetics? Some recent monographs on complexity theory (or complex systems) by other authors give me the impression, that complexity theory is just another name for Synergetics. I hope that the present volume will help to clarify what is meant by complexity or complexity theory and thus will elucidate the relation between these fields and Synergetics.

2. How would you define complexity?

I think it will be hard to find a common ground for the definition of complexity. Whenever we deal with a problem of modern research, in many cases eventually we end up at highly complex questions which can not be answered, probably even not in principle. Thus, in my opinion, at a sufficiently deep level of research, we automatically run into highly difficult and, quite often, unsolvable problems. Complexity may be the borderline, where the transition from simple to nearly (or completely) unsolvable problems happens.

3. What is your favourite aspect / concept of complexity?

My favourite aspect of complexity is the goal of *reducing complexity*. For instance, as in the case of Synergetics, the reduction of complex dynamical phenomena to low dimensional dynamics by means of few concepts, such as order parameters and the slaving principle, etc. The laser may serve as a prototype: Starting from many molecules, many light oscillations and infinitely extended heat baths, finally we have to deal with a single degree of freedom, the laser mode. That this strategy of complexity reduction works has been convincingly demonstrated in the numerous volumes of the Springer Series in Synergetics.

4. In your opinion, what is the most problematic aspect / concept of complexity?

In my opinion, the most problematic aspect/concept of complexity is the fact that it will be very hard to arrive at a common definition of what complexity is. In my opinion, this depends very much on the specific field a scientist works in.

5. How do you see the future of complexity? (including obstacles, dangers, promises, and relations with other areas)

I think a good idea of what the field of complexity is concerned with is provided by the "Complexity Digest", weekly edited by Gottfried Mayer. All the problems and results he quotes are extremely interesting, but cover also an extreme range in which many special disciplines are involved. I think a main goal of human cognition is the *reduction of complexity*. This is absolutely necessary for us to deal with the highly complex problems of our world. My own approach is rather utilitarian. In how far can a future complexity theory help us to better understand the world, especially by reducing complexity? How do general concepts help us to reach this goal? Or must we invoke in the individual cases the highly specialized knowledge of the respective fields of research? Thus, there may be one danger, namely that the whole field becomes so diffuse and diverse that no more any common principles or analogies become visible. It is those common principles that should help us simplify our understanding of the world. Actually, general theories in physics and other fields helped us to subsume the enormous number of individual facts under few basic concepts. I think, my own field "Synergetics" is a modest attempt at reaching the goal of reducing complexity, at least in situations, where qualitative changes occur, namely close to transitions between ordered or disordered states. I think, the main difficulty rests in a proper balance between generality and specification. Probably, this can be solved only in a pragmatic fashion, depending on the taste and skill of the individual researchers. So I think, while it is worthwhile to strive at the exploration of general and common principles, we must be well aware of the limits of such an endeavour.

11

Francis Heylighen

Professor

Vrije Universiteit Brussel, Belgium

---

1. Why did you begin working with complex systems?

I have been interested by all forms of complexity and self-organization since my childhood. I was always a keen observer of nature, being fascinated by complex phenomena such as ants walking apparently randomly across a branch, the cracks that would appear in drying mud, or the frost crystals that would form on grass during winter nights.

As an adolescent, one of my hobbies was keeping aquariums, in which I would try to build a miniature ecosystem complete with soil, plants, invertebrates, and fish. The fish would still need to get food from time to time, and I still had to clean the filter that would collect the dirt they produced, but ideally I would have liked to create a system that is completely autonomous, and is able to sustain itself even in the absence of a caretaker. That would have required more plant life to sustain the food chain, and especially less fish to produce waste products, so it would have made the aquarium less interesting to look at. Therefore, I did compromise in practice. But in my imagination, I was fascinated by what I called "a little world on its own". In my present scientific vocabulary, I would define this idea as a system that is complex and self-organizing to such a degree that it could be viewed almost as a separate, autonomous universe. (Later I discovered a similar idea in the science fiction stories of Stanislaw Lem, a Polish author influenced by cybernetics.)

My fascination for rocks, plants, animals and other phenomena of nature also found an outlet in my early inquiry into the theory of evolution. Like most children nowadays, I had been exposed from an early age to pictures and stories about dinosaurs. The difference, perhaps, is that my grandfather who had collected or drawn these pictures for me was rather scientifically minded, although he was just a primary school teacher. He taught me not only their Latin names, such as Brontosaurus, Triceratops and Tyrannosaurus Rex, but also about the periods in which they lived, and the kinds of creatures that preceded and followed them in the course of natural history. So, from an age of eight or so, I was well aware that life on Earth had evolved, and that plants and animals looked very different in different time periods.

As I became a little older, I started reading introductory books on biology, which explained the mechanism of natural selection behind this evolution. This idea became one of the two fundamental principles on which I have based my scientific worldview. As an adolescent, this mechanism seemed so obvious to me that I was quick to generalize it to other domains, noting that for example ideas and societies also evolved through variation and selection. I called this "the generalized principle of natural selection". Much later, while working on my PhD, I came into contact with other scientists (in particular the great Donald T. Campbell and his disciples Gary Cziko and Mark Bickhard) who had developed a similar philosophy, which they called "selectionism" or "universal selection theory". Its basic assumption is that all complex systems—whether physical, biological, mental or social—have originated through an evolutionary process, which at the deepest level consists of some form of "blind" (not necessarily random) variation, followed by the selective retention of those variants that are most "fit".

In this radical formulation, the theory has few adherents. The reason is that most complexity scientists view Darwin's theory of natural selection with its emphasis on individual organisms or genes as reductionist, ignoring the "whole is greater than the sum of the parts" mantra that characterizes self-organization and complex systems. Yet, I never saw a contradiction between this holistic perspective and my beloved principle of natural selection. The explanation lies in another fundamental idea that I developed while I was 15-16 years old, and which I called the "relational principle".

After reading popular science accounts of Einstein's theory of relativity, I was inclined to conclude, like so many others with a somewhat rebellious streak, that "everything is relative", and that there are no absolute laws nor truths, neither man-made nor natural. (Later I learned that Einstein's own philosophy could hardly have been more different). More recently, this irreverent philosophy has gotten some form of academic respectability under the label of "postmodernism" or "social constructivism". Its main thesis is that different cultures and different people see the same things in different ways, and that there is no absolute criterion to say who is right and who is wrong. But this negative interpretation did not satisfy me: I wanted to truly understand how the world functions.

Therefore, I focused on the positive aspect of relativity: the importance of relations. A phenomenon can only be conceived with respect to, or in relation to, another phenomenon. No phenomenon can exist on its own—without context or environment from which it is distinguished, but to which it is at the same time connected. Later, this idea led me to analyze everything in terms of "bootstrapping" networks, where nodes are defined by their links with other nodes, and links by the nodes they connect. This philosophy is intrinsically holistic: it is impossible to reduce systems to their separate components; it is only through the connections between the components that the system emerges. This relational point of view is not in conflict with selectionism: networks do undergo variation and selection, both at the level of the nodes and links that constitute them and at the higher level of the systems that emerge from clusters of densely linked nodes.

After having formulated the fundamental tenets of my philosophy already while in high school, my challenge was to choose a discipline to study in university. With such a broad interest in complex systems of all types (I had even "reinvented" the concept of social network by drawing a map of all the relationships within my high school class—an exercise that did not make me too popular among my classmates), coupled with a healthy skepticism towards traditional reductionist science, this was not an obvious issue. I hesitated between biology, physics, philosophy and literature, and finally settled on physics, reasoning that I could study the other ones on my own, but with the math underlying physics being so difficult, I would need some solid tutoring if I wanted to become mathematically literate enough to understand the most advanced theories. This reasoning turned out to be correct: studying theoretical physics was hard, but it gave me a basis that allowed me to afterwards investigate a variety of other scientific disciplines on my own.

Within physics, my interest initially did not go towards complexity—which at the time (around 1980) was not yet a fashionable topic. I was lucky enough to get some courses on thermodynamics and statistical mechanics from professors who had worked with the great Ilya Prigogine, the founder of the Brussels School of complex systems. But these particular individuals were less inspiring to me than a young assistant researcher, Diederik Aerts, who was investigating the foundations of quantum mechanics. So, I decided to make, first my Master's thesis, then my PhD on that subject, hoping to be able to elaborate my relational philosophy in a more formal manner. An analysis of the role of the observer in quantum theory together with the creation at our university by Luc Steels of one of the first Artificial Intelligence labs in Europe inspired me to focus on cognition: the processes by which knowledge is acquired and represented.

I submitted a short paper looking at knowledge acquisition as relational self-organization to a conference on cybernetics. There I discovered a whole community of researchers interested in the same transdisciplinary subject of complex systems, their self-organization and cognition. After defending my PhD thesis in 1987, I basically abandoned my work on the foundations of physics, and positioned myself squarely in the field of general systems and cybernetics, hoping that I had finally found my home. Yet, I felt there was still something lacking in that approach, which tended to consider systems as pre-existing, static structures. I missed the evolutionary angle. Therefore I wrote a "Proposal for the creation of a network on complexity research", sketching a theoretical framework that would integrate systems, evolution and cognition.

From the reactions I received, the most interesting one came from a young systems scientist, Cliff Joslyn, who had just developed a similar proposal in collaboration with the veteran cyberneticist Valentin Turchin. They called it the "Principia Cybernetica Project". In 1991 I joined them, and in 1993 I created the project's website. Principia Cybernetica Web (http://pcp.vub.ac.be) was the first and still is one of the most important websites on complex systems, cybernetics, evolution, and related subjects. As such, it has gotten countless students and researchers interested in the domain.

Since then I have been working on integrating these different topics in an encompassing theoretical framework, with a variety of applications in social systems, information technology, psychology and related domains. Independently of our "evolutionary cybernetics" work in Principia Cybernetica, the complex adaptive systems approach had in the meantime become popular, thanks mostly to researchers affiliated with the Santa Fe Institute, such as John Holland and Stuart Kauffman. The similarities between both approaches are much more important than the differences, but there is still enough difference in focus to allow for useful cross-fertilization. It was in part for this purpose that in 2004 I founded the Evolution, Complexity and Cognition (ECCO) research group, which groups most of my PhD students and a number of associate researchers.

2. How would you define complexity?

Anticipating one of the following questions, arguably the most problematic aspect of complexity is its definition. Dozens if not hundreds of authors have proposed definitions, some vague and qualitative, some formal and quantitative, but none of them really satisfactory. The formal ones tend to be much too specific, being applicable only to binary strings or to genomes, but not to complex systems in general. Moreover, even within the extremely simplified universe of binary strings (sequences of 0s and 1s), complexity turns out to be tricky to define. The best definition yet defines the complexity of a string as the length of the shortest possible complete description of it (i.e. the binary program needed to generate the string). However, this implies that a random string would be maximally complex.

The qualitative descriptions can be short and vague, such as "complexity is situated in between order and disorder". More commonly, authors trying to characterize complex systems just provide extensive lists or tables of properties that complex systems have and that distinguish them from simple system. These include items such as: many components or agents, local interactions, non-linear dynamics, emergent properties, self-organization, multiple feedback loops, multiple levels, adapting to its environment, etc. The problem here of course is that the different lists partly overlap, partly differ, and that there is no agreement on what should be included. Moreover, the properties are usually not independent. For example, self-organizing processes normally produce emergent properties, and include feedback loops, which themselves entail non-linearity... Then, not all properties are truly necessary. For example, as I recently noted at a conference where one of such definitions was proposed, a marriage is typically a very complex system that is unpredictable, non-linear, adaptive, etc. Yet it consists of just two agents!

For my own preferred definition, I go back to the Latin root "complexus", which means something like "entangled, entwined, embracing". I interpret this to mean that in order to have a complex, you need two or more distinct components that are connected in such a way that they are difficult to separate. This fits in perfectly with my relational philosophy: it is the relations weaving the parts together that turn the system into a complex, producing emergent properties. To make this qualitative notion more quantitative, I add that a system becomes more complex as the number of distinctions (distinct components, states, or aspects) and the number of relations or connections increases.

The problem with this definition is that it does not lead to a unique number or degree that would allow us to objectively measure how complex a phenomenon is. The reason is that distinctions and connections are not objectively given, easily countable entities: they exist at different levels, in different dimensions, and in different kinds. Aspects can be related to each other across space, across time or across levels. Distinctions can be logical, physical, causal, or perceptual. Adding them all together in order to calculate the overall complexity of a system would be like adding apples and oranges. At best, this definition leads to what in mathematics is called a partial order: X might be more complex than Y, less complex, equally complex, or simply incomparable. It is more complex only if X has all the components and relations that Y has, plus some more.

In spite of this limitation, this definition has some nice characteristics: it is simple and intuitive, and it maps neatly on some of the other simple definitions. For example, complexity, characterized by many distinctions and connections, is situated in between disorder (many distinctions, few or no connections) and order (many connections, few or no distinctions). It also connects the relational and selectionist perspectives: an evolutionary process can be seen as a system of distinctions (variations) and connections (selective continuations) across time. Moreover, evolution generates complexity by increasing variety (number of distinct systems or states) and dependency (systems "fitting" or adapting to each other). I call these twin aspects of complexification: differentiation and integration.

3. What is your favourite aspect / concept of complexity?

As one might have guessed from my biographical notes, I am fascinated by self-organization. Unlike authors like Kauffman, I don't make a strict distinction between self-organization and evolution: both are processes that spontaneously take place in complex systems and that generate more complexity. Evolution tends to be seen in terms of adaptation to an external environment and self-organization as the result of an internal dynamics. Yet, from a systems perspective there is no absolute difference between internal and external: what is internal for the system is generally external for its subsystems. It all depends on where you draw the boundary between system and environment. Thus, as the cybernetician Ashby pointed out long ago, we can view any self-organizing system as a collection of co-evolving or mutually adapting subsystems. Similarly, we can view biological evolution as the self-organization of the ecosystem into a network of mutually adapted species.

I am not just interested in observing self-organization "in the wild", but in creating it in artificial systems. The best-known examples are the computer simulations of organisms, ecosystems and societies that we find in the domains of Artificial Life and Multi-Agent Systems. Such simulations have shown that very simple algorithms (abstract representations of iterative processes) can lead to unexpected complexity, adaptation, and apparently intelligent organization. Let's look at two classic examples.

Genetic algorithms are based on a simple generalization of Darwinian evolution. A variety of potential solutions to a particular problem are generated in the form of strings of symbols. These are tested as to their "fitness", or goodness in tackling the problem. The fittest candidates are selected and made to undergo variation, either by mutation (randomly changing one or a few symbols in the string) or by "sexual recombination" (gluing the first part of one string together with the last part of another). This produces a second generation, which is again selected on the basis of fitness. The best ones of the second generation then reproduce to form a third generation, and so on. After several such generations, the fittest string is typically much better than the ones you started out with, and often produces an elegant solution to a complex problem.

Ant algorithms too are directly inspired by natural self-organization: when ants find food, they leave a trail of pheromones ("smell molecules") along their path back to the nest. Other ants searching for food are more likely to go in a direction where there are more pheromones. If successful, they too will add pheromones, making the trail stronger, and more likely to attract further ants. If no food is found, no pheromones are added and the trail gradually evaporates. In that way, a colony of ants will at first explore their environment randomly, but then gradually develop a complex "roadmap" of trails connecting the nest and the various food sources in the most efficient way.

Applications of self-organization are found not only in computing or in nature, but also in society. Cities, communities, cultures and markets typically emerge through self-organization. Different people with different backgrounds meet by chance, exchange products, services or ideas, thus discovering common interests. This leads to an explicit or implicit collaboration, which is in everybody's interests, and thus binds the assembly of individuals together into a system. The system complexifies as people specialize in certain roles, thus creating a division a labor. This differentiation is counterbalanced by integration, through the creation of communication channels connecting the subsystems together into a larger whole. In that way, a hierarchy of levels is created. Eventually a single individual, such as a president, king, or mayor, may come to occupy the top level, apparently being in charge. But the system is much too complex to be centrally controlled: its "governor" (to use the cybernetic term) may specify high level goals and directions, but the concrete activities are still produced "bottom-up", through the interactions between individuals and subsystems.

Understanding this dynamics allows us to encourage and support it, e.g. when creating new social systems. This happens routinely on the Internet where virtual communities self-assemble around a website or discussion forum. I am particularly interested in the software tools that facilitate such self-organization, and have extensively researched the way they may enhance the "collective intelligence" of the emerging system. Such software tools typically support and guide the interaction between individuals and the information they use, recommending people or resources likely to be useful, and shortcutting the many trial-and-error processes that otherwise would be needed to find an adapted network, e.g. by using an equivalent of ant algorithms.

4. In your opinion, what is the most problematic aspect / concept of complexity?

Conceptually, the most difficult aspect of complexity is still its definition, and the deeper understanding that goes with it. This is probably because complexity requires us to abandon our traditional reductionist perspective, that is to say, our tendency to tackle complex systems by analyzing them into separate components. The opposite perspective of holism, on the other hand, runs the danger of too much vagueness and simplification: just noting that everything is connected to everything else is of little help when tackling concrete problems. The twin principles of relationalism and selectionism, as I sketched them, hold the promise of synthesizing these complementary approaches.

Yet, they still remain quite abstract, and need to be developed into a more concrete and coherent theory.

Practically, the most problematic aspect of complexity is simply coping with it. It is hardly an original observation that our present society is getting more complex every day. The main reason is that modern communication and transport technologies have facilitated interactions between previously remote people, societies or systems, thus increasing their connectivity. Yet, I think that this phenomenon is still insufficiently studied. Indeed, many of our most pressing problems have this growing interdependency at their core.

Let me list some well-known example. Few people nowadays dispute the dangers of global warming. Yet, when it comes to tackling the problem, no one seems to know very well where to start: there are dozens of different possible strategies, from promoting alternative energy to instating a carbon tax, from planting more forests to injecting carbon dioxide into the soil... All of these have different disadvantages and costs attached to them, but—more importantly—they all interact, via their effect on the economy and the ecosystem. This makes the overall effect of any mix of strategies unpredictable. A recently "hot topic" in complexity science is the modeling and detection of terrorist networks. As the world becomes more interdependent, the potential damage created by terrorism grows, yet the terrorist groups become more diffuse and distributed, without a central command that is easy to take out. Finally, the explosive growth of the Internet has brought many benefits, but also created new problems, including information overload and the concomitant stress, cybercrime, and the spread of computer viruses and spam.

The only way we will be able to deal with such dynamic problems is to combat complexity with complexity, i.e. create models and systems based on the same principles of complexity and self-organization as the problem domains they are dealing with. As such, they can co-evolve with the problems, becoming ever better adapted to their moving targets. An illustration of such an approach can be found in the attempts to design a computer security system inspired by the mechanisms of our own immune system. This means that the system would learn to recognize and neutralize computer viruses, worms, intruders, and bugs by the variation and selection of "antibodies" that recognize and disable anything that doesn't behave as it should.

5. How do you see the future of complexity? (including obstacles, dangers, promises, and relations with other areas)

In the longer term, I see some form of complexity science take over the whole of scientific thinking, replacing the still lingering Newtonian paradigm, with its assumptions of separate components, predictable behavior, and static, unchanging laws. However, it is not obvious whether this will be the present, as yet poorly organized, incarnation of complexity science, or some future version that goes under a different name. As the critic John Horgan pointed out, the present complexity wave fits nicely in a sequence of "c-words" that became popular with intervals of about 15 years, but went out of fashion shortly afterwards: Cybernetics, Catastrophe theory, Chaos theory, and now Complexity.

Such ebb and flow of scientific fashions is certainly not limited to complexity. More important than the changes in focus and the accompanying buzzwords, however, is the continuity in the development of the underlying way of thinking. Most complexity researchers would agree that the basic ideas of cybernetics, catastrophe theory and chaos theory still nicely fit under the broad umbrella of complexity science. It is just that we have learned that very specific, and especially mathematical models, such as catastrophes, chaos, fractals, or more recently self-organized criticality, are useful only in a particular, well-defined context, and will need to be complemented by other approaches if we want to apply them to complex systems in general.

The danger is that complexity science would merely become an assortment of advanced modeling techniques that capture with more or less success different aspects of complex systems, but without encompassing theory behind them. I see this danger coming in particular from the remnants of reductionism and determinism that still influence many complexity researchers' way of thinking. Physicists especially have been trained to as much as possible make complete and deterministic models of the phenomena they study, albeit at the cost of studying only relatively simple aspects isolated from their environment or context. This allows them to make more accurate predictions than scientists in, say, biology, medicine or the social sciences, where the subject of investigation cannot be neatly separated out from the things it is connected to.

Now that physicists have started to focus on complexity they tend to take that same attitude with them, applying their impressive array of mathematical tools to the analysis of social, economical or biological systems. While this may produce plenty of interesting insights in the short term, in the long term they need to become aware that it will never provide them with the kind of absolutist "laws of complexity" that many still are looking for. Every complex system has followed its unique evolutionary trajectory and as such is different from any other system. It is only when we become deeply aware of the unlimited number of differences and connections between systems, and the unpredictable evolution this engenders that we will be able to truly build a science of complexity.

Suggested reading

- Ashby, W. R. (1962). Principles of the Self-organizing System. In von Foerster, H. and G. W. Zopf, Jr. (Eds.), *Principles of Self-organization*, Pergamon Press, pp. 255–278.

- Gershenson C. & F. Heylighen (2004). "How can we think the complex?", in: Richardson, Kurt (ed.) Managing the Complex Vol. 1: Philosophy, Theory and Application.(Institute for the Study of Coherence and Emergence/Information Age Publishing)

- Heylighen F. (1997). "Classic Publications on Complex, Evolving Systems: A citation-based survey", *Complexity* 2 (5): 31–36.

- Heylighen F. (2001). "The Science of Self-organization and Adaptivity", in: L. D. Kiel, (ed.) Knowledge Management, Organizational Intelligence and Learning, and Complexity, in: *The Encyclopedia of Life Support Systems* (EOLSS), (Eolss Publishers, Oxford). [http://www.eolss.net]

- Heylighen F., P. Cilliers, & C. Gershenson (2007) "Complexity and Philosophy", in: Jan Bogg and Robert Geyer (editors), *Complexity, Science and Society*, (Radcliffe Publishing, Oxford)

- Holland, J. H. (1996). *Hidden Order: How adaptation builds complexity*, Addison-Wesley.

- Kauffman, S. A. (1995). *At Home in the Universe: The Search for Laws of Self-Organization and Complexity*, Oxford University Press, Oxford.

- Prigogine, I. & Stengers, I. (1984). *Order out of Chaos*, Bantam Books, New York,

- Waldrop, M. M. (1992). *Complexity: The Emerging Science at the Edge of Order and Chaos*, London: Viking.

12

Bernardo A. Huberman

Senior HP Fellow and Director

Social Computing Laboratory, HP Labs, USA

---

1. Why did you begin working with complex systems?

Two strands led to my involvement with complex systems.

The first was related to the work that at the time I was doing in the area of chaos and nonlinear dynamics. While my original involvement with that field came through my expertise in statistical physics and dynamical systems, after a few years of chaos work I realized that I was getting tired of the reductionist approach to physics I had been using in my earlier work, and thus started looking for a connection to phenomena beyond the traditional domains of physics. I had read very early on Phil Anderson's insightful paper "More is Different", and while I found it inspirational I was never able to put his observations into practice on a scale and scope that I found personally satisfying. All of that changed after I started working on chaos. Here was a case where the "more" had been replaced by the "longer" in the sense that one observed incredibly complex patterns unfold in time as a result of very simple deterministic processes.

Through a fortunate collaboration with a psychiatrist at Stanford who was interested in my chaos work, Roy King, I ended up working on a model to explain several symptoms in schizophrenia. This model exploited the then known parameters of dopamine production and reuptake in the brain and generated a number of interesting and observable behaviors. That led to my participation in several conferences populated by neuroscientists and other interdisciplinary people, meetings from which I started to get a sense of what it meant to be able to contribute something meaningful to domains that, while fuzzier to describe, seemed full of interesting phenomena.

Equally important, my work on dopamine dynamics led to a set of conversations and a collaboration with Murray Gell-Mann that was to have a significant impact on my thinking. He was then interested, as always, in a number of phenomena in fields far from physics, including chaos and schizophrenia, and because of my work on dopamine dynamics he asked me to help him with the organization of a special meeting that took place at Les Treilles, a beautiful place in the Var region of France that Anne Gruner-Schlumberger had established for interdisciplinary encounters. During the discussions that led to that meeting Gell-Mann articulated his views of what a study of complex phenomena would entail, views that provided a lucid and exciting description of rather cloudy intuitions that I had about these topics. After the meeting in Les Treilles he went on to establish the Santa Fe Institute, in whose founding workshops I participated. I cannot overstate the importance of what Gell-Mann set out to do with his institute and writings, for he catalyzed a needed movement to break out of mental and institutional barriers that stood in the way of those venturing into this terrain.

The second strand came from the fact that my research was then done at Xerox PARC, where the personal computing revolution was taking place, and I therefore had access to a wonderful infrastructure that was then unique in the world. It did not take me long to realize than rather than using computers to study dynamical problems I could use dynamical systems and statistical mechanics to study distributed computation. To get started in that new domain we designed and implemented an adaptive market mechanism for allocating resources in computer networks, which we called Spawn and that to my surprise actually worked well. Spawn taught me a lot about markets and economics, fields which are paradigmatic examples of complex systems. It also led me to read Fredrik von Hayek's work on economics and emergence, which in spite of its rather dense style, describes a lot of what complexity at the social level is about. And finally, around that time Herb Simon sent me his delightful book "The Sciences of the Artificial", where notions of bottom up complexity, hierarchy and adaptation were wonderfully articulated, so after reading it I decided to make distributed systems, both social and computational, the focus of my next area of research. I haven't regretted it since.

2. How would you define complexity?

There are many definitions of complexity, some of them useful, others insightful and others that have little relevance to most systems. For example, there is the complexity familiar to computer scientists, and which describes whether or not a certain task can be accomplished in a polynomial or exponential time as a function of the size of the problem to be computed. Somehow related to that one is the Kolmogorov measure of complexity, which defines the randomness of a string of numbers by the length of the algorithm that describes it. That notion was eventually extended by Solomonoff to describe the complexity of the laws of nature, and explicated in great detail by Gregory Chaitin. Another quantifiable notion of complexity, which I introduced with Tad Hogg in order to describe hierarchical systems, puts complexity between randomness and ordered systems and is applicable to near decomposable structures, like organizations and complex biochemical processes. And yet another is that found in the dynamics of systems with few degrees of freedom, a complexity characterized by the existence of positive Lyapunov exponents. That positivity leads to chaotic behavior out of very simple deterministic equations and the ensuing complex behaviors one observes when studying their evolution.

But in systems composed of many interacting parts that happen to be intentional (social, economic, ecological) and not necessarily decomposable, complexity arises from the behavior of the whole from the plans of individuals, and thus the panoply of fascinating patterns and apparent regularities we are constantly surprised with.

3. What is your favourite aspect / concept of complexity?

To me, the most interesting aspect of complexity is the ability to predict the behavior of the whole system from the knowledge of individual intentions and dynamics. I guess that it betrays my being a physicist. Think of societies, economic systems and distributed technology, entities that seem to work on a large scale in spite of all the randomness that seems to drive them.

4. In your opinion, what is the most problematic aspect / concept of complexity?

That it is not a science by itself. And yet it is often used in that sense, which is a problem. It leads to abuses by people seeking attention and support for work that does not conform to what my notion of what a scientific field should be.

5. How do you see the future of complexity? (including obstacles, dangers, promises, and relations with other areas)

I'm not very optimistic about the field as such (because we don't have such a field), but very keen on the developments that are taking place in domains that encapsulate the essence of complexity, like economics, social science, systems biology and perhaps more relevantly, the engineering of very large distributed systems.

13

Stuart A. Kauffman

Director

Institute for Biocomplexity and Informatics,
University of Calgary, Canada

---

1. Why did you begin working with complex systems?

In 1964, on my way to medical school, having come from Oxford and
Dartmouth in Philosophy, I took my premedical courses at Berkeley
and found about developmental biology. I had already worked with
theories of neurons turning one another on and off in receptor fields
at Oxford. Jacob and Monod had just shown that genes can turn
one another on and off, for example in the lactose operon, and they
speculated that two genes might repress one another, so that little
circuit could have two states: A on, B off; or B on, A off. So, the same
set of genes could have two different stable patterns of gene expression, which might give you two cell types. That image of alternative
patterns of gene activities corresponding to cell types answered the
question of how the same genome could give rise to multiple cell types.
I fell in love with that.

It was thought at that time that there were around one hundred
thousand genes. Therefore, it was obvious that there was some sort
of regulatory network among the genes. The question that I asked
myself was: Does the genetic network that controls ontogeny has to
be very specific and tuned by evolution, or is there some very large
class of networks with the properties needed for biological systems
and selection only has to do some fine tuning? I did not know any
differential equations at the time, but I knew logic, so that's why I
invented random Boolean networks (RBNs). We can make statistical
ensembles in which we tune the number of inputs per gene and the
choice of Boolean function. As for the generic behaviors, it was obvious
to me that the way to do that was to build networks at random, given
the constraints of the ensemble. I did that when I was in medical

school, and it became the model that I and others have had so much
fun playing with.

2. How would you define complexity?

There are probably eighty definitions of complexity, and any single
definition is going to be inadequate to the task. Part of the definition
is certainly the following: The computer is a tool that is kind of the
opposite to the microscope—it is a macroscope. It allows us to look
at systems with a very large number of parts, where the parts may be
different from one another and may influence one another in different
ways. As we allow them to influence one another, we can ask what is
the collective behavior of the resulting system and whether there are
emergent collective features that we can find. My own passion is to
hope that those emergent collective features will allow us to find laws
that describe the behavior of such systems, that are relatively insensitive to the details of the structure of the system. Robert Laughlin talks
about organizational laws. I think that the order-chaos-criticality that
have emerged in RBNs are examples of these type of laws that are not
reducible to physics.

I think that things get even more mysterious. In my recent book
Reinventing the Sacred, I speak about the amazing arising possibility that we cannot prestate the "adjacent possible" of the biosphere
with respect to Darwinian preadaptations, hence the evolution of the
biosphere. Therefore, the evolution of the biosphere is partially not
describable by natural laws. And if that is right—I stress *if* that
is right—it changes our Western view of the world absolutely dramatically. This is because we have believed since Descartes, Galileo,
and Newton—in our reductionistic worldview—that everything in the
universe is describable by law. If it is true that it is not— then nevertheless Newton's laws get us to the moon and are refined by Einstein
and augmented by the standard model of physics—we have to totally
rethink living in our universe as human beings and what it means to
be a universe that is partially describable by law and partially not.
That brings a variety of questions. What systems can be described
by law? What systems cannot be described by law? What determines
which class you are in? Is there something in between? Right now
nobody knows. Even I do not know whether I am right in my claim,
but everybody seems to think that I am right in my claim that we
cannot prestate Darwinian preadaptations.

Another feature that I am getting very fascinated by is the following: the biosphere—in its partially lawless becoming—has the property that organisms and features of organisms that come to exist at each stage make sense. That is to say, they are selectively useful. So the biosphere is always a self-consistent whole as it evolves to what I call the "adjacent possible" in a non-ergodic universe. And so is the economy as new goods and services come into existence. For example, the channel changer that I talk about in *Reinventing the Sacred*. So the biosphere, the economy, human history, and maybe the universe as a whole, somehow come into existence in a self-consistent but yet partially lawless way, in which what is in the adjacent possible restricts what can become actual. But when the actual happens, like a swim bladder or a channel changer, that in turn changes the actual that changes the adjacent possible in a way that it keeps *becoming* ever in a self-consistent way. And I do not know how to describe that. I do not know if there are any laws to describe it. I do not know what is going on. I am utterly confused. I think it is just wonderful, and it is a part of complexity that does not exist yet. I think this is a new frontier in complexity. Maybe we are talking about a world that *becomes* partially beyond natural law.

3. What is your favourite aspect / concept of complexity?

Before I wrote *Investigations* was the search for laws of organization in complex emergent systems. That is to say, something like Robert Laughlin's organizational laws. For example, in my own work, order, chaos, and criticality in RBNs or the idea of autocatalytic sets, which also cannot be reduced to physics. In the case of networks, it could be scale-free networks. In Laughlin's case, a single iron atom is not rigid but an iron bar is. That would have been my answer. Now, I am so fascinated by this question of partial lawlessness, such that if it is true, it changes what we have thought about science, enlightenment, and Western society. It means that our groundwork, after four hundred years of Descartes, Galileo, and Newton, has become a frontier for me.

4. In your opinion, what is the most problematic aspect / concept of complexity?

I say this with affection. I get worried about some of artificial life. It is one thing to make a model that follows simple rules and gives rise to something that looks like something biological. It is another thing to do the science that says that the rules that your simulation followed have anything to do with real biology. I am not accusing anybody of being naïve. I just think that there is the danger of trying to discriminate when one is making a computer game—that is pretty and interesting but has nothing to do with anything, except being a computer game—and what the criteria are for doing science. That can be problematic, yet there has been great work done in artificial life, too.

5. How do you see the future of complexity? (including obstacles, dangers, promises, and relations with other areas)

Some things simply have become part of how we think. The notion of collective emergent properties in e.g. nonlinear dynamical systems, which was born in the early days of complexity, is now becoming almost a routine pattern of thought in e.g. cell biology—my own area— where it has been resisted for forty years. This is beginning to occur as we begin to understand genetic regulatory networks.

Another area that I think is very important is to think of the cell as an open thermodynamic system, typically non-equilibrium. We have to think about work, power, power efficiency and the notions of what I call propagating organization of process that I describe in *Investigations* and *Reinventing the Sacred*. But we still do not know what we are talking about that is there in what a cell does and what a biosphere does.

Another important area and also one of my own passions is: why is the biosphere so complex? Why do economies get so complex and diverse? I do not think we know, and we need a theory of it.

Another area that is growing is agent-based modeling. This is very interesting, because agent-based models allow one to get into the causal fine structure of a system. An example is a model of the NASDAQ stock market that was made by Vince Darley and Sasha Outkin in *BiosGroup*, a company that I formed. The model is described in a book by Vince and Sasha called *A NASDAQ Market Simulation*. Of course, Chris Langton was among the first to start making agent-based models, as was Josh Epstein and others many years ago. Agent-based models are becoming widely used, and there is an interesting question about how do you relate that to standard mathematics. Agent-based models are algorithmic, and mathematics is formulas, differential equations, partial differential equations, etc. One of the answers is: to find the emergent collective behaviors of the agent-based model and then you try to find ways of writing down effective differential or stochastic differential or partial differential equations for the collective behaviors that you see that emerge in the system.

If I am right about partial lawlessness, I think that understanding what that means for the universe is a huge intellectual task. It may really transform science. But I may be entirely wrong.

We are also beginning to find out in detail how genes actually regulate one another specifically in cells, my own area of passion. And there are wonderful mid-career people, like my friend Joshua Socolar at Duke. Josh is busy trying to make models of genetic circuitry involved in cell fate decision in sea urchin, with about seventy genes or signalling factors. He is a wonderful mix of a very capable physicist who has learned enough biology to really understand what an important problem it is. Or Sui Huang, here at the Institute for Biocomplexity and Informatics, who came from Harvard, who has done wonderful work showing evidence that cell types are high dimensional attractors and that stem cells may differentiate by having pitchfork bifurcations. So, another area of complexity theory is getting tied down very tightly to experiments.

Another area that I think is powerful is in my own area of science: The fact that systems biology—which is what I am busy doing now—invites us to do new kinds of observables, compared to the observables of standard molecular biology. For example, with Matti Nykter, Ilya Shmulevich, and others, we recently published a paper in *PNAS*[1] showing evidence that cells are dynamically critical because in effect the Lyapunov exponent is zero: if you look at neighboring states, they lie on trajectories that neither diverge nor converge. Now, that is not an observable that any cell biologist would have thought about five or ten years ago, and certainly not forty years ago. So new kinds of observables are coming up in biology and other areas, and that is wonderful.

I do not think that there are enormous dangers. I think complexity is emerging as a coherent area of science. Perfectly responsable people are doing it. I think it needs more attention from really good mathematicians, but that will come. I think that the future is very

---

[1]Nykter, M., N. D. Price, M. Aldana, S. A. Ramsey, S. A. Kauffman, L. Hood, O. Yli-Harja & I. Shmulevich (2008), Gene Expression Dynamics in the Macrophage Exhibit Criticality, *Proceedings of the National Academy of Sciences of the USA*, **105**(6):1897-1900.

promising. There are so many things for us to explore, that we are doomed to have a lot of fun.

14

Seth Lloyd

Professor

Massachusetts Institute of Technology, USA

---

1. Why did you begin working with complex systems?

My Ph.D. supervisor at Rockefeller University, Heinz Pagels, came into my office and said,

'OK, Lloyd, we're going to define complexity.'

'But the main feature of complexity is that it resists all attempts to define it,' I replied.

'Bullshit,' he said, 'let's try.'

2. How would you define complexity?

I still think that complexity resists all attempts to define it. I prefer a multifold definition: complexity arises out of a combination of the difficulty of describing or characterizing a system (measured, e.g., in bits), and the difficulty of doing something with the system (measured, e.g., in energy applied/dissipated or in dollars spent). Something is complex if the either or both of these difficulties is great.

3. What is your favourite aspect / concept of complexity?

That everyone knows it when they see it, but no one knows how to define it. And very few people know what to do when confronted with it.

4. In your opinion, what is the most problematic aspect / concept of complexity?

Coming up with an approach to designing / building / operating / interacting with complex systems that is sufficiently broad to be of wide application, and yet still useful.

5. How do you see the future of complexity? (including obstacles, dangers, promises, and relations with other areas)

There will always be a healthy interest in studying and coping with complexity. The fundamental question to my mind is whether we can come up with a unifying theory of complexity that is applicable to all complex systems, or whether the different guises of complexity—physical / biological / social / economic—are too disparate to encompass within a single theory.

15

Gottfried Mayer-Kress

Adjunct Associate Professor

Department of Kinesiology, Pennsylvania State University, USA

Editor

Complexity Digest[1]

---

1. Why did you begin working with complex systems?

In high school we read J.W.v. Goethe's "Faust" about a fellow who sells his soul to the devil to find out "was die Welt im Innersten zusammenhaelt" ("what holds the world together at its innermost levels"). To me that sounded like what high energy particle physics is all about and so I ended up doing my diploma thesis at the largest German particle accelerator, DESY, in theoretical physics on a topic in quantum field theory. After evaluating pages and pages of Feynman diagrams I finally got my diploma, and the question arose whether to continue in the same field with my Ph.D. work.

So one day I asked my adviser where I would be in 30 years, if I was the smartest physicist and I could solve all the technical problems that came up. His answer was that perhaps I could prove quark confinement. Now at that time I was willing to just accept without proof that they are confined, so I asked, "Anything else?" Well, if I was really smart, maybe I can calculate from scratch the mass of the electron, he answered. That didn't sound too exciting either and besides, gluons seemed to be a reasonable explanation of what holds the world together.

As a consequence, I was not so sure, if I wanted to spend the rest of my life as a theoretical particle physicist. It was about that time when I stumbled upon H. Haken's book "Synergetics—An Introduction". It

---

[1]http://www.comdig.org

dawned on me that the ultimate frontiers in physics were not at the very small or the very large, scales far beyond human experience or capability to grasp intuitively, but there was a fundamental frontier just in front of our eyes with phenomena of our daily experience such as the formation of clouds or the growth of slime mold. This new frontier was to the world of complex adaptive systems, chaos, self-organization and other areas that H. Haken and his students had explored under the Synergetics label.

A short time later I was H. Haken's Ph.D. student and he suggested I check into this "new field" called "chaos theory" and he gave me one of the classic papers "Period 3 implies Chaos" by Li and Yorke. I wrote my first paper on some improvements over results by Prof. K. Tomita regarding the invariant measure of the logistic map system. Proudly I took it to a conference and showed it to the great David Ruelle. He glanced at it and then told me that this sort of problems were solved by his students as homework assignments.

Back in Stuttgart I suggested to Prof. Haken to do numerical studies of stochastic perturbations of the logistic system, since the Institute had its own(!) PDP-11/34 mini-computer. My first goal was to have a close look at this chaos that emerges as a consequence of period-3. To my surprise, there was no evidence for chaos at all, only a stable period-3 orbit. Only when I added a tiny amount of noise, the periodic orbit would explode into large-scale chaos. On my next conference I could announce: "Period-3 + Noise Implies Chaos". Later I could classify the kind of chaos observed here as noise-induced intermittency and—where the period-3 orbit disappears in a saddle-node bifurcation—as type-1 intermittency.

During those days I calculated many invariant distributions and Lyapunov exponents, both of them very computer intensive tasks. Whereas most of the jobs of my colleagues took less than 15min, I would sign up for eight hours at a time. Some of it was due to my poor programming skills and one colleague even rewrote my algorithm which sped up the program by more than an order of magnitude. But instead of keeping me away from the computer the result was that I just increased the number of iterations so that the runtime for the jobs remained the same, but the distributions looked much smoother.

It took me quite a while to get permission to run programs over night, because it was institute tradition to turn it off at 5p.m. When we got a three-pen color printer I kept that busy for hours, plotting the first pictures of Julia sets, that later become very popular through the computer graphics of H.O Peitgen and his group.

During my post-doc years at the Center for Nonlinear Studies of the Los Alamos National Laboratory I worked on a number of applications of chaos theory ranging from ocean waves, galaxy distributions, brain dynamics to strategic arms races and the impact on SDI, the Strategic Defense Initiative involving missile defense systems. This last project got the attention of Richard Garwin, who handed our paper around in Washington, DC and—as a result—got me into some trouble back in Los Alamos. But fortunately our model calculations were watertight and at the end the director of the lab came to apologize for any interference with my work.

One day George Cowan, a retired LANL chemist gave a talk about a new institute they wanted to start in Santa Fe with a focus on complexity. During his presentation I noticed that many of the issues they wanted to address were already studied for a number of years in the context of Haken's Synergetics. Nobody seemed to be familiar with that extensive literature (e.g. several dozen volumes in the Springer Series in Synergetics). After I mentioned that to George Cowan and gave him some examples of results from self-organization, etc., he invited me to be one the first post-docs at the new Santa Fe Institute. The first location of the SFI was in an old convent up on Canyon Rd. For lunch we used to walk down to El Farol, a fine old Spanish restaurant, not a bar, which is now legendary for Brian Arthur's "El Farol Bar Problem".

More than in the later, larger venues, this first generation SFI was truly a place for close interdisciplinary interactions and mutual learning from experts in different fields related to complexity. For instance I was always interested in robust or typical behavior of dynamical systems that does not depend on detailed choices for parameters or initial conditions. For example in arms race models it is difficult to measure or estimate parameters such as a "grievance factor" better than 10%. But still, researchers in army labs run spreadsheet simulations of individual runs presenting results with four digit accuracy.

We tried to get a closer grip on producing robust outcomes by running up to tens of millions of simulations of randomly perturbed parameter configurations. Nevertheless, in a high dimensional space even such a large number of parameters gets quickly diluted and one still cannot be sure that the observed results are typical and robust. That is where researchers in complexity theory such as John Holland and Stephanie Forrest have developed tools—such as genetic algorithms—that can be applied to this kind of problems. Instead of just randomly perturbing a reference solution on can "breed" solutions with specific properties. For example if we want to know how big the smallest noise level has to be so that the system has a fair chance to escape from the current type of solutions to a completely different one, then this can be studied by selecting an appropriate "fitness function" and then let a whole population of individual solutions evolve towards the one we want to select for. During the collaboration with Stephanie Forrest on a paper illustrating the application of genetic algorithms to nonlinear dynamical models I learned much about complex systems that go beyond low-dimensional non-linear maps. And this is how and why I got started working on complex systems.

2. How would you define complexity?

Important words such as complexity, life, happiness, physics, etc. are used in many different contexts. On a non-technical level I would prefer to "describe" complexity rather than "define" it because a definition—by definition—always excludes many cases that with some new insight might be included. On a technical level I would say that complexity is a measure that describes the amount of information needed to describe a complex system, assuming that everyone knows what a complex system is. Of course this "definition" is also ambiguous because there are many ways to describe a system. In algorithmic complexity it is necessary to reproduce a sequence of numbers precisely. Therefore an infinite sequence consisting only of 0s has low/zero complexity, because it can be described by "consisting only of 0s". The other extreme is a truly random sequence for which the shortest description is just a copy of the sequence itself, requiring an infinite amount of information.

On the other hand, if we change the meaning of "description" and don't require the reproduction of an individual number, then we can describe a random sequence by "delta correlated Gaussian distribution with mean zero and standard deviation sigma". That means with this "definition" random data also posses zero complexity and the maximum is somewhere in between.

In summary I would say that attempts to formulate a precise general definition can be seen as a philosophical exercise. For practitioners, context dependent, technical definitions with a precise statement of the assumptions is a useful tool to measure an important property of complex systems, namely their complexity.

3. What is your favourite aspect / concept of complexity?

It is the property of complex systems to adapt and learn. Currently I am working on motor learning of humans, how we can learn a complex task through repeated practice. The role of the time between practice sessions and especially the role of sleep is a fascinating open research question.

4. In your opinion, what is the most problematic aspect / concept of complexity?

It is the same that did damage to systems theory, catastrophe theory, chaos and now complexity. That a new and exciting field is over-sold to the lay public, generating unrealistic, almost religious expectations which then backfire and create a bad reputation for the field and researchers who work in it.

5. How do you see the future of complexity? (including obstacles, dangers, promises, and relations with other areas)

Based on what we find in the current literature for Complexity Digest, we see some trends for areas in complexity the have become a focus of research and applications.

Fast progress is made in the area of genetic medicine where it has become increasingly clear that instead of having one gene being responsible for one disease, there can be hundreds of genes involved and their interactions that give rise to the emergence of a specific disease.

Quantum entanglement and its role in quantum information processing is also a very active area of research that has the potential of very powerful new and unexpected applications.

The general interest in complex networks has somewhat leveled-off a bit but it is clear that it provides a foundation for important research in the area of social networks, especially on the Internet, ecological systems with their food networks, and at a cellular level, the signaling pathways within and among cells.

The current global financial crisis shows that existing risk models and others, set-up by math experts or "quants" are by no means sufficient to realistically model and control today's complex, globally connected economic system. Here the application of more sophisticated, complexity based models that also incorporate "common sense" components, together with a deeper, conceptual understanding of the working of complex systems might be able to lead to more stable and robust financial systems.

Maybe the most important future application of complexity involves a self-referential process, where complexity research organizations apply concepts of complexity to their own mode of working. I was always surprised to see organizations such as the Santa Fe Institute and the New England Complex Systems Institute organized and managed in a very traditional way without much application of what we have learned from complexity theory.

In order to solve the most important and urgent global problems such as pollution and global warming scientists will need to self-organize on a global scale. The hope is that a "Global Brain" will emerge that not only applies our collective knowledge to the solutions of these problems but also communicates effectively with governments and the public that it is in the survival interest of individuals and their countries to work towards the common global good instead of competing for local advantages. Today we are far from understanding our Earth ecosystem and their importance for mitigating climate change, especially the role of clouds. Hopefully complexity theory will provide means to stabilize our climate system and avoid maybe one of the biggest catastrophes in human history.

16

Melanie Mitchell

Professor of Computer Science

Portland State University, USA

---

1. Why did you begin working with complex systems?

The short answer is that the phenomena of "emergence" and "adaptation" were (and remain) absolutely fascinating to me. In nature we see many examples of huge numbers of simple elements, interacting with no central control, collectively producing sophisticated adaptive behavior that is far beyond the ability of any single or small group of component elements. Are there any general principles underlying this kind of emergent complex and adaptive behavior? And can we get machines to become intelligent and lifelike using those same principles? These questions are what got me hooked on complex systems.

In the early 1980s, after I graduated from college with a degree in math, I read Doug Hofstadter's book, Gödel, Escher, Bach: an Eternal Golden Braid. That book was my introduction to some of the main ideas of complex systems, It presented a view of the mind as emerging from the brain via the decentralized interactions of large numbers of simple, low-level "agents", analogous to the emergent behavior of cells, ant colonies, and other such systems. I decided that I wanted to study artificial intelligence, and to try to work with Hofstadter on creating intelligent systems based on these ideas.

A year or so later I ended up going to graduate school in computer science at the University of Michigan in order to join Hofstadter's research group. Even at that time, Michigan was a hotbed of work on complex systems. It was the home of the so-called BACH group—named for its original members Arthur Burks, Bob Axelrod, Michael Cohen, and John Holland. These people are pioneers in the field of complexity. I took John Holland's course on "Adaptation in Natural and Artificial Systems", which put forward a view that (in my fellow student Chris Langton's words) "the proper domain of computer science is information processing writ large across all of nature." This really resonated with me. Holland became my co-advisor, along with Hofstadter. These two people have been the most important influences for me in my work.

When I was finishing up my Ph.D. in 1989, Hofstadter was invited to a conference at Los Alamos on "emergent computation". He was too busy to go, so sent me instead, which was a serendipitous opportunity for me. It was at that conference that I met many of the major players in complex systems, and found out about the Santa Fe Institute, a relatively new (at that time) center for research on complexity. After completing my Ph.D. I was invited to spend a summer there, and I was awed both by the beauty of Santa Fe and by the breadth, depth, and novelty of the science that was being done at SFI. I returned to SFI the following summer, and ended up being appointed to the Institute's research faculty. All in all I was there for about eight years, directing SFI's program in adaptive computation.

2. How would you define complexity?

I don't think there is a single good definition, just as there is no single good definition of "self-organization" or "emergence". People use words like this in different ways in different contexts. There is a well-known paper from 2001 by Seth Lloyd describing about 40 different definitions people had proposed, and there have been lots more since. None are really satisfactory, in my opinion.

A more useful approach, I think, is to ask what concepts are most appropriate to employ in characterizing the behavior of these so-called complex systems. My own view is that we will need a combination of concepts from the fields of nonlinear dynamics, information theory, computation, and evolution. People have long made connections among these various fields, but a real interdisciplinary language that captures all these aspects of complexity has not yet been formulated.

3. What is your favourite aspect / concept of complexity?

Given the complexity of "complexity", it's hard to isolate a single "favorite" concept. My current interests are largely focused on the pattern-recognition abilities of complex adaptive systems such as the brain, the immune system, insect colonies, individual cells. and genetic regulatory networks. Pattern recognition, at various levels of abstraction, is one of major activities of all living systems, and life-like pattern recognition has turned out to be rather difficult to capture in computers. I believe there are some general principles underlying the ability of large collections of agents to effect abstract pattern recognition in a decentralized way, and that these principles can help us in designing computer programs with similar pattern-recognition abilities.

4. In your opinion, what is the most problematic aspect / concept of complexity?

The concept of "emergence" is a tough one. It's central to complex systems, yet hard to define. For example, in cognitive science we might say that "concepts" are emergent properties of the activities of networks in the brain. What do we mean, exactly? One definition of emergence is "higher-level global behavior, arising from the collective actions of simple lower-level components, that is more complex than can be achieved by the lower-level components independently". This leaves us to define "higher level", "global behavior", "collective actions", etc. Another definition might be "phenomena we don't yet understand, arising from the collective actions of components we do understand". Again, a rather unsatisfying definition for me, since my intuition is that "emergence" is not just a subjective property, dependent on what we do or don't currently understand.

Many people have tried to define "emergence" formally, but I haven't so far found a definition that is both correct and useful. Emergence is a phenomenon that I believe is "real" in some sense, and is key to understanding complex systems. However I don't think we yet have the conceptual framework or vocabulary to characterize more precisely what this phenomenon is.

5. How do you see the future of complexity? (including obstacles, dangers, promises, and relations with other areas)

I think that, as we increasingly understand complex systems, the concepts and vocabulary we use for describing them will become much more specific, quantifiable, and useful. That is, ill-defined terms such as "emergence", "self-organization", and "complexity" itself will be replaced by new, better-defined terms that reflect increased understanding of the phenomena in question.

One danger is that the field of complex systems might go the way of General Systems Theory or Cybernetics. These earlier disciplines were aimed at answering many of the same questions that complex systems addresses. However, they got a bad name for being, as one Nobel-prize winner described, "well-meant, but premature and intellectually lightweight". It's possible that in 50 years people will similarly criticize early 21st century complexity research.

This is indeed a risk. However, the possible payoffs for pursuing this area are great. In the life sciences, brain science, and social sciences, the more carefully scientists look, the more complex the phenomena are. A good example is the unexpected complexity that is being discovered in genetics and development. New technologies have enabled these discoveries, and what is being discovered is in dire need of a new kind of conceptual vocabulary for describing this complexity and a set of theories about how such complexity comes about and operates. That is something I think complex systems is gradually beginning to offer. As people in the field have joked, we're "waiting for Carnot"— that is, waiting for the right concepts to be formulated to describe what we see in nature. Who knows? It's possible that our Carnot is already among us.

17

Edgar Morin

Emeritus Research Director

Centre National de la Recherche Scientifique, France

---

[1] All my life, I've never been able to be resigned with parcelized knowledge, I've never been able to isolate a studied object from its context, from its antecedents, from its future. I have always aspired to a multidimensional thought. I have never been able to eliminate the internal contradiction. I have always felt that deep truths, antagonists with each other, were for me complementary, without ceasing being antagonists. I never wanted to reduce by force uncertainty and ambiguity.

My definition of complexity gives privileges to:

- Systemic relations.

- Circular causality (retroactive and recursive).

- Dialogic.

- The hologrammatical principle (not only the part is within then whole, but the whole is within the part).

The first definition of complexity cannot provide any elucidation: that which cannot be summarized in a master word is complex, that which cannot be described by a law, that which cannot be reduced to a simple idea. In other words, complexity cannot be summarized by the word complexity, cannot be brought back to a law of complexity, nor be reduced to the idea of complexity. Complexity cannot be something that would be defined in a simple way and would replace simplicity. *Complexity is a problem word, not a solution word.*

---

[1]Translated from the French by Carlos Gershenson

To understand the problem of complexity, it is first necessary to know first that there is a paradigm of simplicity. This is a paradigm that puts order in the universe and chases disorder. Order is reduced to one law, to one principle. Simplicity sees either the One or the Multiple, but it cannot see that the One can be at the same time Multiple. The simplicity principle either separates what is linked (disjunction), or unifies what is diverse (reduction).

It will be necessary to dissipate two illusions that divert spirits from complex thought. The first one is to believe that complexity lead to the elimination of simplicity. Certainly, complexity appears where simplifying thought fails, but it integrates in itself everything that puts order, clarity, distinction, and precision in knowledge. Whereas simplifying thought disintegrates the complexity of reality, complex thought integrates the simplifying modes of thought as much as possible, but refuses the mutilating, reductionistic, unidimensional, and finally blinding consequences of a simplification that is taken for the reflection of what is real in reality. The second illusion is to confuse complexity and completeness. Certainly, the ambition of complex thought is to give account of articulations between disciplinary domains that are broken by the disjunctive thought (which is one of the major aspects of the simplifying thought); this one isolates what it separates, and hides all that connects, interacts, interferes. In this sense, complex thought aspires to a multidimensional knowledge. But it knows from the start that complete knowledge is impossible: one of the axioms of complexity is the impossibility, even in theory, of an omniscience. It endorses the word of Adorno: "the totality is the non-truth". It comprises the recognition of a principle of incompleteness and uncertainty. But it also carries in its principle the recognition of the links between the entities that our thought must necessarily to distinguish, but not isolate from each other.

18

Mark Newman

Professor of Physics

University of Michigan, USA

---

1. Why did you begin working with complex systems?

As a graduate student in England (a "postgraduate" as the British say), I worked in conventional theoretical physics, but as a postdoctoral researcher at Cornell University I drifted away from the field of my doctoral work and became interested in the mathematical work that theoretical biologists, physicists, and mathematicians were doing on biological systems, especially in evolutionary biology and ecology. Reading more about the subject I came across the work of the Santa Fe Institute, which particularly interested me because the approaches its researchers were taking felt familiar to me as a physicist, but the problems were much broader than those I'd been brought up on and to me more interesting.

So when my position at Cornell came to an end I accepted a postdoctoral job at SFI, later becoming resident faculty there and spending a total of six very enjoyable years in Santa Fe. While at the Institute, I learned a great deal about the broader world of complex systems research. With a continuous stream of internationally known researchers coming through and giving talks on the widest imaginable range of topics, from physics and biology to politics and sociology, I heard about all sorts of fields and approaches that I'd not previously known of. My own research during that period covered a wide range of topics as well, including mathematical biology, paleontology, and sociology, in addition to conventional physics. By the time I left in 2002 to take a position at the University of Michigan, I had, I suppose, become a complex systems researcher, albeit one whose approach is still firmly rooted in physics. I feel lucky to have become involved with the subject quite early in its development and to have had the chance to meet and learn from many of the greats of complex systems research.

2. How would you define complexity?

"Complexity" is an ill-defined term. There are, famously, dozens of definitions in circulation and no consensus about which are reasonable and which are not. So I avoid using the word. If I have to describe what I do I say that I work on "complex systems". On the face of it, that might appear to be just as vague a term as "complexity", but in practice there is a much better consensus about its meaning. Commonly a complex system is defined to be a system composed of many components or "agents", which interact with one another so that the system as a whole is more than just the sum of its parts. We say the system shows "emergent behaviours", collective responses of all the parts that make it up. An example would be a market, such as a stock market, which, among other things, sets the prices of a wide range of items by a process of bargaining between traders. No single trader in a market can say what the day's price for something will be, but once traders start trading with one another the price "emerges". Other frequently cited examples of complex systems include the Internet, in which the interactions of many people and their computers produces outcomes and behaviours that were rarely anticipated before computers were connected together, and more generally human society itself, in which the interaction of billions of human beings produces a rich variety of social phenomena. Animal societies too can show interesting emergent behaviours—social insects such as ants, for example, are commonly studied as an example of a complex system.

Most of these systems are also "complex" in the everyday sense of the word—far too complex for us to understand their every detail. Current research in complex systems therefore tends to focus only on one or two aspects of a system at a time. Researchers working in mathematical finance for instance might make computer models of a stock market with only one stock and simplified trading rules that can be understood more easily than a real market with its bewildering range of financial instruments, rules, inputs, and regulations. Researchers working on human societies might focus on an aspect such as how people decide who to vote for, or how they decide what to buy, ignoring most of the myriad complexities that affect our real lives. Nonetheless, even these much simplified approaches have given us a huge amount of understanding about human behaviour and life on Earth, things that just a few years ago were thought to be outside the range of exact science.

3. What is your favourite aspect / concept of complexity?

My own work focuses on networks—the patterns of connections between agents in a complex system. I find these patterns intriguing and, for the moment at least, they are a clear favourite of mine. Take the pattern of connections between human beings in a society, for example: who is friends with whom, or who works with whom. Though such patterns have been studied for decades, their study has received a big boost with the advent of cheap computers for data collection, and particularly with the appearance of on-line social networks that provide new insight into the ways that people communicate. With all the new data pouring in, we are beginning to develop for the first time a really clear picture of the large-scale organization of human society. We see how larger communities—the inhabitants of a town or the students in a school—split into smaller ones—circles of friends or coworkers—and how those split into still smaller groups, and so on. We see the diversity of patterns of interaction, with some individuals having huge numbers of casual acquaintances while others have just a small number of close friends. The famous phenomenon of the "six degrees of separation", in which (almost) any individual can be connected to any other by a short chain of acquaintances, perhaps only about half-a-dozen steps long, has now been observed in all sorts networks from the collaborative networks of business people and scientists to the networks of who calls whom on their mobile phones. And all of these phenomena can be linked to real-world outcomes such as the spread of fads or fashions, or the speed with which this year's flu moves through the population. As the volume of data available to us steadily increases and we develop new ways to analyse it and understand what it's telling us, I anticipate a lot more discoveries in this area in the near future.

4. In your opinion, what is the most problematic aspect / concept of complexity?

This is a difficult question: there are lots of problematic aspects in every area of science, and complex systems is no exception. But for me perhaps the most problematic aspect is the comparison of the predictions of science against real-world data. In many cases scientists make mathematical or computer models to explain the workings of a particular complex system. Then they compare the predictions of their model against experimental data to see if model and reality agree. In many cases they do, but is this enough to say that the model is right? The answer, of course, is no. To say it was would be like saying, "Bears like honey; my wife likes honey; therefore my wife is a bear." Just because one model gives a correct prediction about something doesn't mean that another model, perhaps completely different, could not also give the same prediction. To some extent this is a problem with all of science, but it is a particularly difficult one when studying systems that are truly "complex". The complexity of a market or an ecosystem or a human community is often so great that it can be hard to say when one's model has captured the important points and when it has not.

A related problem is choosing the level at which one should model a complex system. Models of the stock market, for example, range all the way from simple mathematical models embodied in just a single equation, to massive computer simulations designed to predict all the fluctuations and details of real trading. Both extremes have their uses, but for most scientists the sweet-spot lies somewhere in between, and finding it is one of the most difficult—and contentious—parts of the job.

5. How do you see the future of complexity? (including obstacles, dangers, promises, and relations with other areas)

We have taken great strides in the last 20 years or so, and progress is coming at an ever quicker pace, so there is every reason to hope for fascinating and important progress in the future. On the other hand, complex systems still seems like a fragmentary subject area rather than a unified whole. Current work is divided among a wide variety of practitioners working in different areas, some commercial, some academic, and using all sorts of different techniques. Although I don't believe there will ever be a unified science of complex systems—there will, and should be, always a good range of different approaches—I think that the field could greatly benefit if researchers in different areas were more aware of the methods and tools have have been developed by others. I do see some progress in this direction—I've met computer scientists who know all about biology and physicists who know about sociology (and vice versa)—and as long as this trend continues I think we will see truly new approaches to complex systems science in the not too distant future.

19

Grégoire Nicolis

Professor

Interdisciplinary Center for Nonlinear Phenomena
and Complex Systems
Université Libre de Bruxelles, Belgium

1. Why did you begin working with complex systems?

My acquaintance with systems composed of interacting subunits—of which complex systems constitute a particularly significant class—dates from my research in nonequilibrium statistical mechanics during my PhD and post doctoral years at the Université Libre de Bruxelles under the direction of Ilya Prigogine and the University of Chicago under the direction of Stuart Rice. The principal questions of interest at that time were linear transport theory and the derivation of kinetic equations describing the approach to equilibrium, using perturbative expansions. Things began to shift in 1966 and onwards. The extension of thermodynamics to open systems far away from equilibrium and the discovery of nonequilibrium induced self-organization phenomena, the dissipative structures, by the Brussels group to which I had the privilege to participate, showed that systems of interacting particles are capable of exhibiting unexpected behaviors not reducible to those of their individual elements that would be next to impossible under equilibrium conditions.

There was a lot of excitement accompanying these discoveries and a real sense of urgency to elucidate the onset and the principal characteristics of the associated phenomena. During the 1970's to mid 1980's this program was tackled successfully both at the macroscopic and the microscopic levels of description using the methods of nonlinear dynamics and chaos theory on the one side, and those of the theory of stochastic processes (master and Langevin equations) on the other. Theoretical work in fields ranging from fluid mechanics, optics, material science and chemistry to biology in conjunction with laboratory scale experimental developments showed that similar behaviors were recurring in very different contexts. Meanwhile starting in the 1980's new issues were being raised in connection with life sciences, large scale natural systems such as the atmosphere, and human systems such as competing agents in the stock market where the elementary subunits are no longer particles but entities capable of reasoning, of reacting and of adapting. The key question of interest here was prediction, to which I became exposed through my long standing collaboration with my wife Catherine Nicolis.

Eventually, it was the search for a unified description of such problems that gave rise to the idea of complex systems as a field of science in its own right, embodying, in fact, the most exciting and the most innovative facet of systems composed of interacting particles. Throughout my work in the field, I have been stressing the view of complexity as part of fundamental mathematical and natural science, and the need to keep investing on mathematically and physically motivated issues and on the sharpening and further development of the associated techniques. There can be no "soft" approach to complexity: observing, analyzing, modeling, predicting and controlling complex systems can only be achieved through the time-honored approach provided by "hard" science. The novelty brought by complex systems is that in this endeavor the goals as we set them in traditional approaches are reformulated and the ways to achieve them are reinvented in a most unexpected way. This view has been at variance with the one that had prevailed for some time, namely, that the perception of a system as complex reflects essentially the practical difficulty to gather detailed information, following the presence of often prohibitively large numbers of parameters and variables masking the underlying regularities. Fortunately this latter view which, if true, would identify "complexity" to "complication" and would reduce it to nice metaphor and an appealing way of putting things, is now recognized to be obsolete.

2. How would you define complexity?

Complexity is the conjunction of several properties (some of which are reviewed in the sequel) and, because of this, no single formal definition doing justice to its multiple facets and manifestations can be given. It is useful to compare this with the concept of nonlinearity (itself a necessary condition for complexity) which in contrast can be defined straightforwardly, as it corresponds to a structural feature built in the evolution laws, namely, deviation from strict proportionality between the effects and the underlying causes.

One popular idea surrounding complexity is that an object can be regarded as complex when there is no short description of it. The concept of algorithmic complexity pioneered by Andrei Kolmogorov and Gregory Chaitin has the great merit to propose a quantitative measure capturing this idea: the complexity of an object in its digitalized expression of binary sequence of length K is the size of the shortest computer program (measured in number of bits) generating it. Although algorithmic complexity accounts for certain features of natural complex systems in its basic philosophy it is fundamentally different from the complexity one is concerned with in nature, where one seeks to identify emergent properties, concerted behavior and evolution. In particular, algorithmic complexity is insensitive to the time needed to accomplish a program (assuming that the latter will eventually halt). But in nature it is important to produce certain forms of complexity as the system of interest evolves in real time. The probability to produce a prescribed pattern/sequence out of the enormous number of a priori possible ones is usually exceedingly small. In contrast, under appropriate conditions dynamical systems are capable of exploring their state space continuously thereby creating information and complexity; at the same time they act like efficient selectors that reject the vast majority of possible patterns/sequences and keep only those compatible with the underlying dynamics. Furthermore, dissipation allows for the existence of attractors that have asymptotic stability and thus reproducibility. It therefore seems legitimate to state that algorithmic complexity is a static, equilibrium like concept whereas physical complexity takes its full significance in a dynamic, nonequilibrium context. To tackle physical complexity, one needs a nonequilibrium generalization of classical information theory.

Attempts at a compact definition—or at least measure—of physical complexity beyond its algorithmic aspects as formalized by the Kolmogorov-Chaitin complexity have been reported in the literature. A interesting measure, on the grounds of its relation to prediction, is the amount of information necessary to estimate optimally conditional probabilities. In quite a different vein, one associates complexity to "value" of some sort, for instance, the time required to actually retrieve a message from its minimal algorithmic prescription. In this view a message is complex, or deep, if it is implausible and can only be brought to light as a result of a long calculation. This introduces the time element that is so conspicuously absent in the

Kolmogorov-Chaitin complexity. While capturing certain aspects of physical complexity, none of these definitions/measures manages to fully encompass its multiple facets. The question, how to define complexity is thus likely to remain open for some time to come. It may even turn out to be an ill-posed one: after all as stressed already above, complexity does not reflect any built-in, immediately recognizable, structure as is e.g. the case of nonlinearity; it is, rather, a set of attributes that spring into life from the laws of nature when the appropriate conditions are met.

3. What is your favourite aspect / concept of complexity?

A most appealing aspect of complexity research is to provide a forum for the exchange of information and ideas of an unprecedented diversity cutting across scientific disciplines, from pure mathematics to biology to finance. On the one side one witnesses the encounter and cross-fertilization of nonlinear dynamics, chaos theory, statistical physics, information and probability theories, data analysis and numerical simulation, in close synergy with experiment. And on the other side, insights from the practitioner confronted with large scale systems as encountered in nature, technology or society, many of them outside the strict realm of traditional mathematical and natural science, where issues eliciting the idea of complexity show up in a most urgent manner, are increasingly integrated into the general framework. This multilevel approach, with its conjunction of complementary views and its reassessment of principles and practices confers to complex systems research a marked added value beyond the traditional disciplinary approach to the understanding of nature.

It is often stated that fundamental science is tantamount to the exploration of the very small and the very large. This assertion becomes, simply, obsolete in the light of complex systems research. There exist huge classes of phenomena of the utmost importance, fundamental as well as practical, between these two extremes waiting to be explored in which the system and the observer—the external world and ourselves—co-evolve on comparable time and space scales. This adds further credence to the relevance and unique status of complexity in contemporary science.

Coming now to more concrete issues, one aspect that I view as especially innovative is that complex systems lie at the cross roads of the deterministic and probabilistic views of nature. Let me make this point more precise. The conjunction of multiplicity of possible outcomes, of the sensitivity associated with occurrence of criticalities or of deterministic chaos and of the lack of a universal and exhaustive classification of all possible evolution scenarios characteristic of complex systems, confers to them an intrinsic randomness that cannot be fully accounted for by the traditional deterministic description, in which one focusses on the detailed pointwise evolution of individual trajectories. The probabilistic description offers the natural alternative. The evolution of the relevant variables takes here a form where the values featured in a macroscopic, coarse grained description are modulated by the random fluctuations generated by the dynamics prevailing at a finer level. This highlights further the variety of the behaviors available and entails that the probability distribution functions, rather than the variables themselves, become now the principle quantities of interest. They obey evolution equations like the master equation or the Fokker-Planck equation which are linear and guarantee (under mild conditions on the associated evolution operators) uniqueness and stability, contrary to the deterministic description which is nonlinear and generates multiplicity and instability.

Thanks to its inherent linearity and stability, the probabilistic description of complex systems is the starting point of a new approach to the problem of prediction, in which emphasis is placed on the future occurrence of events conditioned by the states prevailing at a certain time as provided by experimental data. This approach finds nowadays intensive use in, among others, operational weather forecasting, where it is known as ensemble forecasting.

A second appealing aspect at the very basis, in fact, of complexity is the emergence of levels of description obeying to their own laws. There is an apparent paradox accompanying the transition to complexity. On the one side, complexity seems to follow its own rules reflecting the emergence, at some level of description, of new qualitative properties not amenable to those of the individual subunits. But on the other side, since the laws of nature are deterministic, these properties are bound to be deducible from the interactions between lower order hierarchical levels. Because of this, the concept of emergence is still viewed by many as an expression of ignorance.

A first instance where this apparent conflict can be resolved thereby allowing one to quantify the concept of emergence and to establish a connection between different hierarchical levels pertains to the macroscopic description, in which individual variability and more generally deviations from a globally averaged behavior are discarded. Suppose that the system of interest is described by a set of n macroscopic observables, where n can be as large as desired and that it operates in the vicinity of a criticality. An important result of nonlinear dynamics is that for certain (generic!) types of criticalities there exist a limited number of collective variables, to which one refers as order parameters, obeying universal evolution laws characteristic of the criticality at hand, to which one refers as normal forms. All other variables follow passively the evolution of the order parameters. The specific nature of the original evolution laws is immaterial as long as it gives rise to the relevant bifurcation and enters only to specify the values of the parameters present in the normal form. We here have a first instance of how a new level of description following its own rules is being generated. Notice that the essential property sought here is closure, namely the existence of an autonomous set of laws for the relevant variable pertaining to the level of description considered.

A most exciting point is that under certain (generic!) conditions the probabilistic description itself acquires the status of an emergent property, free of heuristic approximations, starting from a deterministic microscopic level description. This passage from the Liouville equation to the master or Fokker-Planck equations depends crucially on the unstable, chaotic character of the microscopic dynamics. A second important ingredient is a judicious choice of "states", through an adequate partition of the full phase space spanned by the variables descriptive of the elementary subunits into cells. As the microscopic trajectory unfolds in phase space transitions between cells—states— are induced, which are isomorphic to a probabilistic process. Such considerations are also instrumental for building a microscopic theory of irreversibility, also viewed in this context as an emergent property.

It should be realized that there are limits to the hierarchical view, reflecting the failure of the decoupling between levels of description. This is what happens, in particular, in nanoscale systems, in systems subjected to strong geometric or nonequilibrium constraints, or in phenomena such as earthquakes, floods, and financial crises associated with the occurrence of extreme values of the relevant variables. A full scale description becomes then necessary, in which the fine details of the structure of the probability distributions begin to matter. Universal laws governing some key observables can still be extracted, examples of which are given by fluctuation type or more generally large deviation type theorems.

4. In your opinion, what is the most problematic aspect / concept of complexity?

To remain relevant, complexity research needs to strike the right balance between the search for generic features and qualitative insights (which should anyway remain one of its main goals) and the specificities and hard empirical facts that are as a rule present in any concrete system of interest.

Complexity has to consolidate further its roots as part of fundamental science and, at the same time, demonstrate its identity and its specificity as compared to related disciplines like nonlinear dynamics and statistical physics. And it has to produce results that could not be otherwise obtained, useful to the practitioner, in major fields outside the traditional realm of mathematical and natural science like brain research, the economy, or the evolutionary and adaptive behaviors in which current attempts, though promising, still remain in their infancy.

Challenges of this kind are usually not part of the goal of traditional scientific disciplines, from cosmology and the nanosciences to molecular biology and sociology. They reflect the special status claimed by complexity research as well as its ambition to constitute a new, "post-Newtonian" scientific paradigm and to play an integrating role in today's highly fragmented scientific landscape.

5. How do you see the future of complexity? (including obstacles, dangers, promises, and relations with other areas)

Complexity is a fundamental discipline in its own right. It contributes to our understanding of nature and has the potential of a significant impact on science and technology. It opens new perspectives, proposes novel strategies and addresses long-standing problems and real world issues of relevance in everyday life. On these grounds, it should be expected to play an increasingly important role in the future.

Complexity research attracts audiences of an unprecedented diversity. There is an inherent danger that this might eventually prove to be a "mixed blessing" in view of the highly heterogeneous, if not loose, character of the community. It is not rare to even see the concept of complexity grossly misused. Care should thus be taken to improve the often poor communication currently existing between different subgroups. In particular, natural and social sciences should come closer together and share expertise.

To achieve the goal of shaping a coherent, clearly recognizable supercritical size, complexity community appropriate training and collaborative programs will have to be initiated. There is at present a lack of complexity related education in most academic institutions,—even though the subject appeals to young people—and a lack of public awareness of the benefits of the complex systems approach. This gap should be filled, and this will probably require imagining and implementing practices of a new kind. In doing so one should not succumb to the temptation of dilution, encouraging prospective complexity researchers to learn a "little bit" of "everything". On the contrary, a hard core of researchers of high level technical expertise in fundamental aspects and in the elaboration of advanced methodologies should be secured. This knowledge should become available in appropriate forms to less technically oriented parts of the community through joint ventures of various kinds. And conversely, insights from these sectors should be integrated to stimulate new developments at the fundamental and technical levels.

There is at present a strong academic community of high level researchers worldwide working in complexity related topics (even though some of them would rather stress their disciplinary identity), as well as a number of case studies where complexity research has been successful. They should constitute the nucleus from which the above envisioned activities could be successfully materialized.

Suggested reading

- G. Chaitin, *Meta Maths*, Atlantic Books (2006).

- P. Gaspard, *Chaos, Scattering and Statistical Mechanics*, Cambridge University Press, Cambridge (1998).

- P. Glansdorff and I. Prigogine, *Thermodynamic theory of sturcture, stability and fluctuations*, Wiley, London (1971).

- H. Haken, *Synergetics*, Springer, Berlin (1977).

- S. Kauffman, *The origins of order*, Oxford University Press, New York (1993).

- G. Nicolis and I. Prigogine, *Self-organization in nonequilibrium systems*, Wiley, New York (1977).

- G. Nicolis and C. Nicolis, *Foundations of complex systems*, World Scientific, Singapore (2007).

20

Jordan B. Pollack

Professor of Computer Science

Brandeis University, USA

---

1. Why did you begin working with complex systems?

I was interested in psychologically plausible Natural Language Processing in graduate school at Illinois and re-invented a kind of neural network to do coordinated decisions among syntax, semantics, and lexical choices, using spreading activation and lateral inhibition. This lead to more formal work as connectionism rose from its previous ashes, and I focused on the fundamental capacity of recurrent neural nets to represent compositional information and process formal languages.

I stumbled across curious phenomenon in these recurrent networks. I trained my network from a finite set of examples of a handful of 7 regular languages, and then tried to relate the network behavior to finite state automata that can recognize those languages. The first idea I had was to look at all possible states and then cluster states together which were within a small value, ϵ, of each other. This worked for a few of the languages, but for others, as I reduced ϵ, the number of clusters would rise exponentially. I thought the network was broken, but then a couple of years later, I realized that this phenomena was like Mandelbrot's paradox that measuring a coastline with shorter and shorter rulers would make the coast longer and longer. So we then plotted all the states of a recurrent network to see what it looked like, which made some interesting pictures, one of which, the "magic mushroom" became the mascot of my lab at Ohio State University.

At first we called this a "strange automata" after the terminology of the "strange attractor" which was being used in complex systems research. But in fact, I had inadvertently created an analogy between automata with a finite state memory, and the recurrent neural network as a dynamical system which could theoretically at least have infinite memory in real numbers. The training algorithm somehow drove the weights of the network to have so many real states, but instead of filling the space, they were located on some kind of fractal geometric form. I found an isomorphism between the higher order recurrent networks that I was using, and the "Iterated function Systems" (IFS) promoted by Michael Barnsley in his book "Fractals Everywhere" as a radical solution to the image compression problem.

This led to a crazy new idea: In AI and Cognitive Science, recursive thoughts-within-thoughts, and sentences-within-sentences indicate generative capacity of a linear-bounded Turing machine, and also justify the recursive programs and data-structures of LISP as valid models for mental processes. The lack of neural plausibility of AI programs always annoyed me. Now, here was a valid alternative: Maybe the recursive elements of mind were simply the natural side-effect of non-linear brain dynamics! The fractal mind emerges from the chaotic brain! I started working on the implications and requirements for demonstrating dynamical cognition and presented it at the first NIPS conference (where I remember being laughed out of the room.)

With the fractal mindset in place, when trying to understand the capacity of recursive auto associative memory (RAAM) we found that the decoding part of the RAAM was also an IFS. We were able to show that under certain mathematical conditions, RAAM can represent infinite sets of trees as equivalence classes of transients to a fractal attractor.

The big payoff in dynamical cognition that I envisioned was a new kind of fractal reconstructive memory. Imagine that instead of storing a story or life experience in bits in a file, the firing pattern of relevant populations of neurons was a fractal (due to plant-like wiring). Then, like a Mandelbrot set determined by a few parameters, a complex memory could be stored as a few parameters which could later reconstruct that complex pattern of experience. The parameters for a fractal model, in other words, could be a new kind of symbol that contains, rather than points to, a memory to be reconstructed on demand.

At the time, Barnsley and his associates were claiming amazing results from fractal image compression while holding their algorithms as trade secrets. Yuval Fisher published very clear algorithms that involved brute force search for tiling images with reductions from a population of larger tiles from the same image. It was awfully slow! I assigned our own "fractal inverse" problem of finding weights for neural networks to reconstruct fractal attractor patterns to several graduate students and Postdoc's, without great success. In the meantime, some prominent philosophers, like Tim Van Gelder and Terry Horgan who had taken a hold of the dynamical cognition hypothesis, wrote so many papers about it. Unlike me, they didn't actually have to implement a working replacement for powerful symbolic AI systems. I realized that many parameters would have to be set, and started working on evolution of complexity as I moved to Brandeis University.

2. How would you define complexity?

Well, the mind is complex! Once I had access to supercomputers, I realized the AI problem was NOT going to be solved by faster and faster computers under Moore's law. Fortran programs are still running on supercomputers and are about the same size they've always been, but are just consuming more floating point operations. Bigger computers didn't lead to more complex and sophisticated software. It just led to bloat.

In order to explain this realization, I defined "biological complexity" informally as an estimate of the number of "unique moving parts" or non-redundant lines of code necessary to formally specify a complex biological system, like the brain, immune system, cell metabolism, etc. I flatly stated that this could require a billion to 10 billion lines of code.

Computers are big enough and reliable enough to run such programs, but software engineering has reached some kind of practical limit, where the largest programs were 10-100 millions lines of code. Moreover, these legacy software systems were always breaking, and needed hundreds of programmers just to keep them working. This practical limit to software complexity as the roadblock for mechanical intelligence can be understood by analogy to mechanical flight. As one scales up from a paper model airplane to a vehicle capable of carrying people, the forces involved grow non-linearly. Trying to balance a plane in turbulent high winds by having the passengers lean left and right is a recipe for crashing. What the Wright brothers solved was the scaling up of control through the aileron principle, using wing shape (or later flaps) to dynamically balance the plane. My opinion is that to achieve biologically complex systems—like cognition—we need some kind of machine learning that could scale up software construction beyond what teams of human programmers could write or maintain.

As I worked on this new approach based on understanding and simulating co-evolution in nature and in robotics and problem-solving, I've searched for a definition of complexity which captures enough elements of design complexity in nature to be used to measure artifacts. I've considered moving parts, repetition/reuse, elegance, underlying generativity, and lost-work in selectionism. Of course, there are many existing approaches to defining complexity, the main two being

1. Chomsky's hierarchy of generative complexity in infinite sets of strings (languages) and the related automata models which are necessary to generate them, and

2. Kolmogorov-Solomonoff algorithmic complexity, the idea that a string is only as complex as the smallest algorithm on a universal computer which could produce it.

The elegance and beauty of the complexity hierarchy of generative formal languages consumed many early careers in computer science, and had practical applications in programming language compilers, but has not proven influential in understanding biological complexity (other than L-systems) other than human language.

Kolmogorov complexity is intuitively satisfying because a uniform or patterned string, like an all-white or all-black or a checkerboard image can be generated by a tiny program. It is unsatisfying because it ultimately asserts that a random number is the most complex object, because the shortest program would be a print statement with the whole number. This definition captures no internal structural complexity as we see in nature at all!

Starting from the Kolmogorov model, consider the space of computer programs that can draw into an image buffer or onto a graphics display screen. On the one extreme, a simple program can generate a set of random pixels. Other very simple programs might draw lines, polygons, or circles, and depending on the set of primitive subroutines to include filling, gradients, patterns, sines and cosines, programs can start to draw more complex canvases.

On the other extreme a mythical AI might be able to pass the "Artistic Turing test" and create paintings that fool humans into buying them as art. There actually was an early AI program called Aaron, which drew saleable art using a flat bed plotter. We would expect that a program would get "larger" as the drawings became more complex and sophisticated—in keeping with Kolmogorov's measurement.

Then there is the Mandelbrot set, which is considered the mother of all fractals. A very small program generates what appears to the eye to be tremendously complex and often beautiful images, depending on the coloring algorithm. Zooming into the set finds self-similar shapes, and all kinds of fantastic psychedelic imagery.

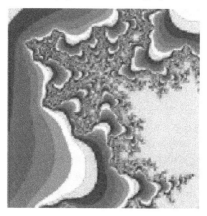

Cover and Hart's book on Information Theory, which actually uses a picture from the Mandelbrot set on the cover, succinctly describes the algorithm as:

"For different points c in the complex plane, one calculates the number of iterations of the map $z_{n+1} = z_n^2 + c$

(starting with $z_0 = 0$) needed for to cross a particular threshold. The point c is then colored according to the number of iterations needed."

And they go on to say that "The fractal is an example of an object which looks very complex but is essentially very simple. Its Kolmogorov complexity is essentially 0." (p. 471).

So it seems that looks can be deceiving, when a very complex picture is the result of a small program. Charles Bennett has defined Logical Depth, as a variation of Kolmogorov complexity that also considers the amount of CPU time needed to generate the string.

Another important notion has been called contamination, that when strings are filtered or selected by a more complex or intelligent system, they are contaminated by that complexity, so they may not be as simple as they first appear. For example, when a man-made object, like a cigarette butt, is found among the sand and rocks on a beach, it may be a simple roll of cotton in paper, but because it is actually the product of a sophisticated manufacturing effort, it is more complex than it appears.

This notion allows disagreement with Cover and Hart: A beautiful image from the Mandelbrot set looks simple because the program is small. But if a human used a fractal program repeatedly to search for a beautiful image from the Mandelbrot set to put on the cover of their book, they have selected it from an infinite set of pictures using their own aesthetic sense, so it is art, non-zero in complexity.

The notion that selection adds to the complexity of an object should involve both the haystack principle—how big was the population from which the sample was selected, and the IQ principle—how intelligent was the selection operator?

When studying the idea of programs which can draw more and more artistic images or compose music, I came across many models of "creativity" which work analogously to evolutionary algorithms, with a generator and a selector. The generator might generate many alternative pictures (from what set?), and the selector chooses among these pictures (using what criteria?) for the best one. We defined the Pablo Picasso Paradox as the question: Where does the IQ for this creativity reside?

Consider for a generator, the simple program which sets each pixel to a random color. Generate 1 million such artworks. Then consider as the selector, a million-sided die, which randomly selects one of the artworks. This is of course garbage in and garbage out. But if Pablo

Picasso generated hundreds of images, and a random die was used to select one, it would be art. Furthermore, and perhaps counter-intuitively, if Pablo had the time to look through the millions of random dot images and found one he thought was beautiful, it would also be art!

The paradox of such models of using creativity to generate complexity (or art) is that the IQ must to be in the generator, or the selector, or both. Putting a human in the loop of defining representations, fitness functions, or acting as the blind-watchmaker only contaminates the results. The key to evolving complexity is to evolve the intelligence in both the generator and selector.

Finally, I want to mention an idea I call "Platonic Density." What is the difference between the simple computer program that computes a pseudo-random color or draws a circle, and another simple one that draws a visually complex picture from the Mandelbrot set?

Maybe such creative programs can be thought of as "inter-dimensional scoops" which retrieve information from somewhere in the immaterial (Platonic) universe, and instantiate that information, perhaps imperfectly, in the material universe, e.g. computer display memory or physical state or shape. The circle drawn on a raster display (or instantiated as a plate or Frisbee) is never a perfect circle, because is irrational. The Mandelbrot set can only be drawn to the resolution of your computer display or color printer.

A program that creates complexity through dissipating CPU time is a pointer to an address in the Platonic universe, and so the information located there has some intrinsic density, higher for the Mandelbrot set program than the random number program. Platonic Density is the ratio of observer-perceived complexity to the amount of computational work to retrieve it.

Both the development of complexity in nature and human creativity can be viewed as scoops that have developed more IQ as to where to find patterns with high Platonic Density.

3. What is your favourite aspect / concept of complexity?

My favorite aspect of complexity is that—once you get past the floating point numbers—it unites with computing to form a whole new world view based on the evolution of dynamical states of information. This computational view has been developing for 50 years, from ideas that the universe could be built on cellular automata, to a broader base in all the sciences dependent on automated data collection and

mining, to changes in the new economy based on information trans-
actions, as well as the emergence of new media independent of paper,
plastic, or film.

I feel that computer science as it is currently defined is at a bit
of a dead end, but that informatics more broadly construed must
take its place as a "school" of thought, along with schools of Science,
Humanities, Theology, and Law, based on Information rather than
reduction, subjectivity, divinity, or precedence, respectively.

4. In your opinion, what is the most problematic aspect / concept of complexity?

It is taking too long a time to filter knowledge out to the mainstream
community and educational system.

5. How do you see the future of complexity? (including obstacles, dangers, promises, and relations with other areas)

As we achieve more and more automatic design of complex systems,
there are dangers in misappropriation, crossed expectations, and im-
penetrability. Misappropriation of science and technologies to create
weaponry, especially when countermeasures are inexpensive, is a waste
of resources. Crossed expectations that systems are smarter or more
human than they are can lead to frustration by users. And Impenetra-
bility is already a problem in the government and large institutions,
and if complex systems begin to be adapted, rules may be enforced
for no reason other than "we can't change the system."

Finally, to the extent that complexity solves the intelligent robot
problem as a synthetically biologically complex electronic life-form,
there is a great promise that human civilization could exist on a much
smaller carbon footprint, using robotics instead of third-world slave
economies, and getting economies of scale without consumption of
cheap products of mass production. However, should we move to a
situation where the human population could and should drop by 90%
to e.g. reverse global warming, we might not want the robots to decide
which 90%!

21

Peter Schuster

Head

Institute of Theoretical Chemistry, University of Vienna, Austria

President

Austrian Academy of Sciences

---

1. Why did you begin working with complex systems?

I started to work with complex chemical reaction networks in the late nineteen sixties, long before the analysis of complex systems became a topic of general interest in science. The specific systems we were working with were the reaction networks of *in vitro* evolution.

2. How would you define complexity?

Complexity is defined best by negation: Simple systems are characterized by simple causality that manifests itself in easily recognized input-output relations, e.g. 'more input leads to more output' or 'more input leads to less output', simple systems have no inherent limits of predictability in the sense that only uncontrollable inputs lead to unpredictable outputs, and simple systems are commonly low-dimensional in the sense that a few independent parameters are sufficient to describe the system exhaustively. In other words, complex systems have sophisticated input-output relations, are inherently unpredictable and high-dimensional.

3. What is your favourite aspect / concept of complexity?

The beauty of complex systems is the enormous variety in dynamical behaviour ranging from multiple states and hysteresis to spontaneous pattern formation, oscillations and deterministic chaos.

4. In your opinion, what is the most problematic aspect / concept of complexity?

The unpredictability of complex systems that have severe consequences for society and/or environment will be even more problematic in the future.

5. How do you see the future of complexity? (including obstacles, dangers, promises, and relations with other areas)

Complexity research has been a fashion for more than two decades by now. Attempts to find a common method for the analysis of complex systems failed so far and I believe that only a powerful 'toolbox' to handle complex systems will be available in the future. Complex systems, like nonlinear systems are 'individuals' and escape naïve classification. Complex systems research will turn out to be an important issue in science as well as in economics and social sciences. Teaching complex systems, now done at certain places only, will become routine at universities. At present, complexity research is already integrated into several individual disciplines like physics and economics and this will be even more so in the future. Nevertheless, there will be also room and necessity for the interdisciplinary approach.

22

Ricard V. Solé

Research Professor

Complex Systems Lab, Universitat Pompeu Fabra, Spain

External Professor

Santa Fe Institute, USA

1. Why did you begin working with complex systems?

As a youngster, at some point I started to read about fossils and evolution but I was also fascinated by mathematics. So I became a strange kind of theory-inclined naturalist. I used to spend long times in the forests and started to see regularities (unfortunately already known by others). Later on, molecular and cell biology became the center of my attention and I decided to become a biologist. Then, two chance events made me look towards complexity. One was a physics teacher who opened my eyes to the beauty and power of physical sciences. I wanted to become a physicist too and eventually I did both degrees. That was a great combination—as I discovered soon— to go deep into complex systems and during those years I was able to start thinking in some of the key problems of complexity. The other was a spanish translation of the proceedings of a famous conference, entiled "Towards a theoretical biology". A bunch of fascinating papers exploring gene networks, development, evolution or brain dynamics captured my imagination and interest forever.

As soon as I completed my degrees, I became involved in a PhD on models in developmental biology, but again an accident changed my path. At some point, I read an article entitled "Chaos" in Scientific American (by Crutchfield, Farmer, Packard and Shaw, the famous Santa Cruz team) and I became trapped by nonlinear dynamics, strange attractors and the like. I changed my PhD topic and completed a new one on spatiotemporal chaos and criticality in evolutionary ecology. As soon as I finished my PhD, I recruited four people

as PhD students and—in the middle of a rather hostile environment—
the Complex Systems Lab came to life. During those years I met (and
collaborated with) several key players in the field, including Stuart
Kauffman, Brian Goodwin and Per Bak and eventually came to the
Santa Fe Institute, where I found the perfect place to develop complexity in a really interdisciplinary community.

2. How would you define complexity?

I think there is a more or less robust consensus in relation to what
makes a complex system: some set of elements interacting in such
a way that higher-order, system properties emerge. These properties
cannot be reduced to the properties displayed by individual parts and
thus some kind of "irreducible order" is at work. I would add to this
tentative definition that complexity is strongly tied to an unbreakable
dialog between system and parts: the global pattern is generated by
individuals, who also receive feedback from the system. Moreover, it
seems clear that the global pattern can be explained (at least in most
cases) by completely ignoring most of the individual details, and that
seems also inextricably linked to complex behavior.

3. What is your favourite aspect / concept of complexity?

I have been always interested in how complexity has emerged through
evolution. Why there are complex life forms instead of (just) simple replicating cells? The major transitions involved the emergence of
some new class of qualitative form of organization linked to a new
way of manipulating information. The french biologist François Jacob
said that "our organism is a sort of machine for predicting the future:
an automatic forecasting apparatus". I think the picture of a complex
organism as a system performing computations and being able to cope
with environmental change is at the core of the problem of how complexity emerges. In my view, one of the key problems is understanding
how information is integrated and how this triggers the emergence of
computation.

Closely related to this question is the universality of the process: is
there any relation between the ways information/computation are performed in living systems versus man-made artifacts? We know some
of the answer: biology works by means of extensive re-use of previous
components whereas engineers can use any kind of building blocks (at

least in principle). Moreover, evolution cannot foresee the final result of its dynamical paths, whereas the engineer designs with a purpose. Nevertheless, we have found that in some designed systems the constraints acting on the dynamics strongly influence system's evolution, eventually forcing individuals (engineers) to perform some class of extensive tinkering. The result is a network that displays life-like patterns. Is this an indication of some class of convergent evolution shared by nature and technology?

4. In your opinion, what is the most problematic aspect / concept of complexity?

There is a sociological aspect: complexity buzz words have become widespread beyond the academia and entered the world of politics, social sciences and journalism (not to mention "holistic" groups). The misuse of terms such as "chaos" or "emergence" is common and to some extent damages the credibility of complex systems sciences. Given the increasing importance of interdisciplinary approaches to our future (climate change, biodiversity or new technologies) this should be corrected. The general public should be able to get clear information on why complex systems approaches are necessary and not get the impression that complexity is just a bunch of metaphors.

5. How do you see the future of complexity? (including obstacles, dangers, promises, and relations with other areas)

I think complexity has been already successful. After some years of being negatively seen as a too broad perspective, its basic nature is being rapidly incorporated to multiple disciplines. As it happened with other fields (such as chaos, fractals or neural networks) most key concepts will become part of mainstream science. Some pieces of the field, such as allometric scaling theory, will stand as perfect examples of the power of complexity in a given area of knowledge and others are slowly percolating within some hard terrains such as economy. The success of systems biology (the old theoretical biology under a new label) is a clear indication of the relevance of system-level views to rapidly developing areas where huge amounts of data need to be understood. Both theoretical and practical ramifications of current knowledge of complex systems are percolating through all domains of science and I think this is great news.

23

Tamás Vicsek

Professor of Physics

Department of Biological Physics. Eötvös Loránd University, Budapest, Hungary.

---

1. Why did you begin working with complex systems?

In 1980 I started working on percolation theory by mainly carrying out computer simulations of the interesting variants of the original model. This was then a flourishing branch of statistical physics and also suiting well my related education. Doing research on percolation gradually evolved into investigating fractals, since the incipient percolating cluster at the threshold probability is a self-similar geometrical object. Fractals and multifractals are already complex enough, but I made a big step towards complexity studies when my friend Prof. Mitsugu Matsushita showed me the fractal bacterial colony he had produced in his lab in 1990.

Over the years I have been stepping up on the complexity staircase: first studying many interacting bacteria, then making a model for flocking in the presence of perturbations and finally, eventually arriving at considering the group behaviour of people. Thus, in a way I was guided by the idea of universality while trying to locate the same patterns of behaviour in more and more complex systems.

2. How would you define complexity?

The world is made of many highly interconnected parts over many scales, whose interactions result in a complex behaviour needing separate interpretation for each level. This realization forces us to appreciate that new features emerge as one goes from one scale to another, so it follows that the science of complexity is about revealing the principles governing the ways by which these new properties appear.

In the past, mankind has learned to understand reality through simplification and analysis. Some important simple systems are successful idealizations or primitive models of particular real situations, for example, a perfect sphere rolling down on an absolutely smooth slope in vacuum. This is the world of Newtonian mechanics, and involves ignoring a huge number of simultaneously acting other factors. Although it might sometimes not matter if details such as the billions of atoms dancing inside the sphere's material are ignored, in other cases reductionism may lead to incorrect conclusions. In complex systems, we accept that processes occurring simultaneously on different scales or levels matter, and the intricate behaviour of the whole system depends on its units in a non-trivial way. Here, the description of the behaviour of the whole system requires a qualitatively new theory, because the laws describing its behaviour are qualitatively different from those describing its units.

Knowledge of the physics of elementary particles is therefore useless for these higher scales. Entering a new level or scale is accompanied by new, emerging laws governing it. When creating life, nature acknowledged the existence of these levels by spontaneously separating them as molecules, macromolecules, cells, organisms, species and societies. The big question is whether there is a unified theory for the ways elements of a system organize themselves to produce a behaviour typical for wide classes of systems. Interesting principles have been proposed, including self-organization, simultaneous existence of many degrees of freedom, self-adaptation, rugged energy landscape and scaling (for example power-law dependence) of the parameters and the underlying network of connections. Physicists are learning how to build relatively simple models producing complicated behaviour, while those working on inherently very complex systems (biologists or economists, say) are uncovering the ways their infinitely complicated subjects can be interpreted in terms of interacting, well-defined units (such as proteins).

To give a definition would be too ambitious for me at this stage. I would rather give a short list of signatures or various aspects of complex systems. It seems that complex systems are i) hierarchical, ii) have many degrees of freedom, iii) tend to exist near an edge (fragile optimized state) and, as a further identification criterion I would also mention: iv) we do not have a satisfactory theory describing them in general...

3. What is your favourite aspect / concept of complexity?

What I like about complexity the most is that it occurs all the time around me in the most common everyday life as well as in subjects of standard research. I do not mind at all that for a moment I feel I almost understand it, and then I have to realize how far I am from finding a good solution to its description in scientific terms.

The most provoking aspect is the hierarchical nature of a complex system. What is "hierarchical" anyway? We have no satisfactory answer even to this simple question (an answer which allows a general formalism applicable to all kinds of systems with features associated with various kinds of hierarchy).

4. In your opinion, what is the most problematic aspect / concept of complexity?

Complexity has become a popular buzzword used in the hope of gaining attention or funding—institutes and research networks associated with complex systems grow like mushrooms.

Why and how did it happen that this vague notion has become a central motif in modern science? Is it only a fashion, a kind of sociological phenomenon, or is it a sign of a changing paradigm of our perception of the laws of nature and of the approaches required to understand them? Because virtually every real system is inherently extremely complicated, to say that a system is complex is almost an empty statement—couldn't an Institute of Complex Systems just as well be called an Institute for Almost Everything?

5. How do you see the future of complexity? (including obstacles, dangers, promises, and relations with other areas)

What we are witnessing in this context is a change of paradigm in attempts to understand our world as we realize that the laws of the whole cannot be deduced by digging deeper into the details. In a way, this change has been invoked by development of instruments. Traditionally, improved microscopes or bigger telescopes are built to understand better particular problems. But computers have allowed new ways of learning. By directly modelling a system made of many units, one can observe, manipulate and understand the behaviour of the whole system much better than before, as in networks of model neurons and virtual auctions by intelligent agents, for example. In this sense, a computer is a tool improving not our sight (as in the microscope or telescope), but our insight into mechanisms within a complex system. Further, computers are used to store, generate and analyse huge databases—the fingerprints of systems that people otherwise could not comprehend.

Many scientists implicitly assume that the understanding of a particular phenomenon is achieved if a (computer) model provides results in good agreement with the observations and leads to correct predictions. Yet such models allow us to simulate systems far more complex than those that have solvable equations. In the Newtonian world, exact or accurate solutions of the equations of motion provide predictions for future events. As a rule, models of complex systems result in a probabilistic prediction of behaviour, and the form in which conclusions are made is less rigorous compared to classical quantitative theories, even involving elements of 'poetry'.

At present I believe that the best approach to the characterization of the main features of complex systems is the investigation of networks most of them (through great simplifications) may be mapped into. For example, understanding the structure of protein interaction networks is likely to lead us closer to uncovering the secret of the little but efficient units of life, e.g. cells.

24

Stephen Wolfram

Founder & CEO

Wolfram Research, Inc., USA

---

1. Why did you begin working with complex systems?

It is a slightly complex story. I started working in physics when I was an early teenager. Mostly I worked on particle physics, but I also thought a lot about the foundations of thermodynamics and statistical physics. And around 1978 I got very interested in the question of how complex structure arises in the universe—from galaxies on down.

Soon thereafter, I became very involved in computer language design—creating a precursor of what is now *Mathematica*—and was very struck by the process of going from the simple primitives in a good language, to all of the rich and complex things that can be created from it.

In 1981 I felt like taking a little break from my activities in doing physics, computing, and starting a company. I decided to do something "fun". I thought I would look back at my old interests in structure formation. I realized that there was the same central question in lots and lots of fields: given fairly simple underlying components, how does such-and-such a complex structure or phenomenon arise?

I decided to try to tackle that question—as kind of an abstract scientific question. I think what I did was very informed by my experience in creating computer languages. I tried to find the very simplest possible primitives—and see what happened with those. I ended up studying cellular automata, and using those, discovered what I thought were some pretty fundamental facts about how complexity can arise.

2. How would you define complexity?

Formal definitions can get all tied up in knots—just like formal definitions of almost anything fundamental: life, energy, mathematics, etc. But the intuitive notion is fairly clear: things seem complex if we do not have a simple way to describe them.

The remarkable scientific fact is that there may be a simple underlying rule for something—even though the thing itself seems to us complex. I found this very clearly with simple cellular automata. And I have found it since with practically every kind of system I can define. And although they were not really recognized as such, examples of this had been seen in mathematics for thousands of years: even though their definitions are simple, the digits of things like the square root of 2 or pi, once produced, seem completely random.

I might say that sometimes our notions of complexity end up being very close to randomness, and sometimes not. Typically, randomness is characterized by our inability to predict or compress the data associated with something. But for some purposes, perfect randomness may seem to us quite "simple"; after all, it is easy to make many kinds of statistical predictions about it. In that case, we tend to say that things are "truly complex" when the actual features we care about are ones we ca not predict or compress.

This can be an interesting distinction—but when it comes to cellular automata or other systems in the computational universe, it tends not to be particularly critical. It tends to be more about different models of the observer—or different characterization of what one is measuring about a system—than about the fundamental capabilities of the system itself.

3. What is your favourite aspect / concept of complexity?

That it is so easy to find in the computational universe.

One used to think that to make something complex, one would have to go to a lot of trouble. That one would have to come up with all sorts of complicated rules, and so on. But what we have found by just sampling the universe of simple programs is that nothing like that is the case. Instead, it is really very easy to get complexity. It is just that our existing science and mathematics developed in such a way that we avoided looking at it.

The ubiquity of complexity has tremendous consequences for the future of science, technology and in a sense our whole world view.

4. In your opinion, what is the most problematic aspect / con-

cept of complexity?

In the early 1980s I was very excited about what I had discovered about the origins of complexity, and I realized there was a whole "science of complexity" that could be built. I made quite an effort to promote "complex systems research" (I would have immediately called it "complexity theory", but wanted to avoid the field of theoretical computer science that was then using that name).

But it is always a challenge to inject new ideas and methods. People liked the concept of complexity that I had outlined, and increasingly used it as a label. But I was a bit disappointed that the basic science did not seem to be advancing. It seemed like people were just taking whatever methods they already knew, and applying them to different systems (usually with rather little success), and saying they were studying "complexity".

It seemed to me that to really study the core phenomena—the true basic science of complexity—one needed a new kind of science. So I ended up spending a decade filling in that vision in my book *A New Kind of Science*. I am happy to say that since my book appeared, there has been an increasingly good understanding of the new kind of basic science that can be done. There has been more and more "pure NKS" done—that gives us great raw material to study both the basic phenomenon of complexity, and its applications in lots of fields.

5. How do you see the future of complexity? (including obstacles, dangers, promises, and relations with other areas)

It is already underway... but in the years and decades to come we are going to see a fundamental change in the approach to both science and technology. We are going to see much simpler underlying systems and rules, with much more complex behavior, all over the place.

Sometimes we are going to see "off the shelf" systems being used—specific systems that have already been studied in the basic science that has been done. And often we are going to see systems being used that were found "on demand" by doing explicit searches of the computational universe.

In science, our explorations of the computational universe have greatly expanded the range of models that are available for us to use. And we have realized that rich, complex, behavior that we see can potentially be generated by models that are simple enough that we can realistically just explicitly search for them in the computational universe.

In technology, we are used to the standard approach to engineering: to the idea that humans have to create systems one step at a time, in a sense always understanding each step. What we have now realized is that it is possible to find great technology just by "mining" the computational universe. There are lots and lots of systems out there—often defined by very simple programs—that we can see do very rich and complex things.

In the past, we have been used to creating some of our technology just by picking up things in nature—say magnets or cotton or liquid crystals—then figuring out how to use them for our purposes. The same is possible on a much larger scale with the abstract systems in the computational universe. For example, in building *Mathematica*, we are increasingly using algorithms that were "mined" from the computational universe, rather than being explicitly constructed step-by-step by a human.

I think we are going to see a huge explosion of technology "mined" from the computational universe. It is all going to depend on the crucial fundamental scientific fact that even very simple programs can make complexity. And the result is that in time, "complexity" will be all around us—not only in nature, but also in the technology we create.

When I started working on complexity nearly 30 years ago, the intuition was that complexity was a rare and difficult thing to get. In the future, everyone will be so exposed from an early age to technology that is based on complexity that all those ideas that seem so hard for people to grasp now will become absolutely commonplace—and taken for granted.

Epilogue

Given the diversity of answers to the five questions on complexity, it is difficult to summarize them in all their richness. In other words, the answers are not *reducible*, as relevant information would be lost. I will limit myself to discuss the similarities and tendencies of most of the answers.

Complexity practitioners come from almost every field. Still, many techniques were developed initially in physics, so there is an abundance of complexity researchers with a physics background. Also, most researchers use computer simulations. Thus, even when computer scientists are not dominant in complexity research, practitioners have gained immensely from the development and use of computers as tools for exploration.

Almost everybody agrees that there is no agreement in a definition of complexity. The "problem" is that it is a very general concept, so its precise meaning changes from context to context. It is quite debatable whether complexity is a science, a theory, or a field. This depends not only on the difficulty of defining complexity, but on the different notions people have of the concepts 'science', 'theory', and 'field'. Nevertheless, very few would say that the study of complexity is not *scientific*. Moreover, it has been used for engineering purposes as a *method*. Finally, some people also see complexity as a new *paradigm*, as opposed to the reductionist approach that has dominated science since Newton.

Most contributors had their own particular favorite aspect or concept of complexity. Many of them liked its generality (it can be applied across scales and fields), creativity (it can generate new patterns), ubiquity (it is everywhere around us), or beauty. Properties of complex systems, such as self-organization, emergence, adaptation, and evolution were also mentioned. I believe this is a clear indicator of their relevance to the study of complex systems.

The most common problematic aspect of complexity was the misuse of the concept. This might be a consequence of its generality, which leads to a lack of a precise definition. It will be important to distinguish clearly the scientific study of complexity, so that this does not become contaminated from the uses of the word in non scientific arenas and the problems this might carry back into science.

As for the future of complexity, most contributors are optimistic. Certainly, we will know for sure only when it comes. But we can expect further advancements in the study and understanding of complex systems, whether they carry the label 'complexity' or not. Will it become the dominant scientific paradigm, replacing the old Newtonian assumptions? Will it be the basis of a new worldview that will enable us to cope better with our complex world? Only time will tell.

About the Editor

Carlos Gershenson (b. 1978) is a Mexican researcher affiliated with the New England Complex Systems Institute and the Vrije Universiteit Brussel. He defended a PhD at the Center Leo Apostel for Interdisciplinary Studies of the Vrije Universiteit Brussel, on the "Design and Control of Self-organizing Systems". He holds an MSc degree in Evolutionary and Adaptive Systems from the University of Sussex, and a BEng degree in Computer Engineering from the Fundación Arturo Rosenblueth, México. He also studied five semesters of Philosophy at UNAM, México.

He has a wide variety of academic interests, including complexity, self-organizing systems, artificial life, evolution, cognition, artificial societies, and philosophy. He has been an active researcher since 1997, working at the Chemistry Institute, UNAM, México, and a summer at the Weizmann Institute of Science, Israel. He has about fifty scientific publications in books, journals, and conference proceedings. He co-edited the book "Worldviews, Science, and Us: Philosophy and Complexity". He is a contributing editor of Complexity Digest (www.comdig.com), and book review editor for the journal Artificial Life.

About Complexity

This volume consists of short, interview-style contributions by leading figures in the field of complexity, based on five questions. The
answers trace their personal experience and expose their views on the
definition, aspects, problems and future of complexity.

The aim of the book is to bring together the opinions of researchers
with different backgrounds on the emerging study of complex systems.
In this way, we will see similarities and differences, agreements and
debates among the approaches of different schools.

Index